# Titan Unveiled

Titan Unveiled

Titan Unveiled

# Titan Unveiled

## Saturn's Mysterious Moon Explored

RALPH LORENZ AND JACQUELINE MITTON

Princeton University Press    *Princeton and Oxford*

Copyright © 2008 by Princeton University Press

Published by Princeton University Press, 41 William Street,
    Princeton, New Jersey 08540

In the United Kingdom: Princeton University Press, 3 Market Place,
    Woodstock, Oxfordshire OX20 1SY

ISBN: 978-0-691-12587-9

Library of Congress Control Number: 2007938922

British Library Cataloging-in-Publication Data is available

This book has been composed in Granjon, Berthold Akzidenz Grotesk Condensed

Printed on acid-free paper. ∞

press.princeton.edu

Printed in the United States of America

10  9  8  7  6  5  4  3  2  1

# Contents

# List of Illustrations
# and Tables

ILLUSTRATIONS

CHAPTER 7

## LIST OF TABLES

# Preface

This book describes the most recent episodes in the unfolding story of the exploration of Saturn's largest moon, Titan, arguably a world that offers planetary scientists a richer scientific bounty than any other in the solar system, apart from Earth.

*Titan Unveiled* is a successor to our earlier book, *Lifting Titan's Veil* (Cambridge University Press, 2002), written before the arrival of the Cassini–Huygens mission. However, it is by no means necessary to read *Lifting Titan's Veil* before reading the present book. Each forms a self-consistent account, and emphasizes different aspects of Titan and its exploration.

As in our earlier book, first-person anecdotes have been corralled into short sections headed "Ralph's Log" (a device suggested by Simon Mitton in 1998 when, it should be noted, blogging was virtually unheard of). These parts are overtly from personal experience but the main narrative is necessarily subjective too.

This is not a deliberate strategy to emphasize the contributions to Titan research of the first author, but rather reflects two factors. Firstly, the surface of Titan and its interaction with the atmosphere have been the most mysterious, and continue to be perhaps the most interesting, aspects of Titan; Lorenz has been intimately involved in two projects hat have so far shed the most light on these aspects: the Huygens probe and the RADAR instrument on the Cassini orbiter. The book, therefore, gives prominence to these investigations. Other Cassini scientists, if writing a similar account, might reasonably highlight other aspects of Titan (and we would encourage them to do so), but we believe that the atmosphere and surface in particular will interest general readers most.

Second, as with *Lifting Titan*'s Veil, we wanted to illustrate a couple of aspects of planetary science, and indeed, of science in general. Planetary science encompasses a broad range of disciplines, and the complexity of Titan as a planetary object means that studying it embraces a large number of them. This diversity of topics and techniques has been challenging, educational and fun to explore, and we wish to convey some of that excitement and variety. Not all the findings or endeavors we describe represent successes. This too is a feature of science as a process. A theory that best fits the facts of the time is often found wanting as new data emerge, and sometimes an old observation upon which an elegant theory is founded is eventually determined to be false. Thus, science is principally about process, not conclusions; it does not lead unerringly to some ultimate truth, but rather gropes around, making false steps in its search for ever more accurate and succinct descriptions and explanations of the universe around us.

We hope that you, the reader, will enjoy our account and learn from it, and that you will consider some subjective emphasis a reasonable price to pay for our report being "from the front line."

Although we name a number of individuals throughout the narrative of our book, the international Cassini–Huygens project, and the exploration of Titan as a whole, result from the efforts of literally thousands of people, all of whom can claim credit for their contributions to this great adventure. We are grateful to them all.

RALPH LORENZ
*Columbia, MD, USA*

JACQUELINE MITTON
*Cambridge, UK*

# Titan Unveiled

Titan Unveiled

Titan Unveiled

# 1. The Lure of Titan

On July 1, 2004, the *Cassini* spacecraft arrived at Saturn after a journey from Earth lasting almost seven years. At 6.8 m in length, this monstrous robotic explorer was the largest western spacecraft ever to be dispatched on an interplanetary mission. Its battery of scientific instruments was designed to return images and data not only from the giant planet itself and its spectacular ring system, but also from members of Saturn's family of over fifty moons. Foremost in interest among the diverse collection of icy worlds in orbit around Saturn was Titan, a body so special, so intriguing in its own right that *Cassini* carried with it a detachable package of instruments—named the *Huygens* probe—that would parachute through Titan's atmosphere to observe its surface.

By any reckoning, Titan is an unusual moon. It is 5,150 km across—nearly 50 percent bigger than our own Moon and 6 percent larger than Mercury. If it happened to orbit around the Sun, its size and character would easily make it as much a planet as Mercury, Venus, Earth, and Mars. But the landscape of this extraordinary world remained hidden to us throughout the first decades of the space age, partially because of Titan's remote location and partially because it is swathed in a thick and visually impenetrable blanket of haze. Thanks to *Cassini–Huygens* and the technological advances that have vastly extended the reach of ground-based telescopes, the situation has now changed dramatically. Titan is undergoing an all-out scientific assault both by the most powerful telescopes on Earth and from the cameras and radar aboard *Cassini*, the flagship international space mission. This observational barrage, topped off by the *Huygens* probe's daring drop down to the surface of Titan, is serving to unveil this enigmatic moon, revealing more of its intriguing features than we have ever seen before.


## THE IMPERATIVE TO EXPLORE

When the two *Voyager* craft sped past Jupiter and Saturn between 1979 and 1981, they returned a wealth of new information about the two giant planets and their moons. But the images and data received from these missions were essentially snapshots—fleeting opportunistic glances at worlds demanding more serious and systematic attention. And as far as Titan was concerned, the results of these flybys were especially disappointing.

Observing Titan was a high priority for the planners of the *Voyager* missions, and in November 1980, *Voyager 1* passed Titan at a distance of 4,394 km. The encounter sent the spacecraft hurtling out of the plane of the solar system and prevented it from exploring any more moons or planets. However, curiosity about Titan was so great that the sacrifice was considered worth making.

A principal reason for the great interest in Titan was the fact that it possesses a significant atmosphere. Astronomers had been aware of Titan's atmosphere since 1944, when Gerard Kuiper announced that spectra he had taken of Titan revealed the presence of methane gas. Therefore, planetary scientists were not going to be surprised to find haze or clouds in the atmosphere, but at the very least, they hoped that parts of Titan's surface would be visible when *Voyager* arrived.

Unfortunately, those hopes were completely dashed. The whole of Titan proved to be shrouded from pole to pole in opaque orange haze. *Voyager*'s camera was sensitive only to visible light, and the spacecraft carried no instruments (such as an infrared camera or imaging radar) capable of probing below the haze. *Voyager* was able to return some important new data about the atmosphere but virtually nothing about the surface.

The exploration of the Jupiter and Saturn systems continued to beckon, however, and the next logical step was to send orbiters to make close and detailed observations over a sustained period of time.

Between the two of them, Jupiter and Saturn possess five of the seven largest moons of the solar system, and they both have far more known

Figure 1.01.   A *Voyager*-era montage of the Saturnian system. The sizes reflect the quality of imaging obtained on each object, rather than their actual size. Because little detail could be seen on Titan, it was perhaps unfairly portrayed small, at the upper right. (NASA)

moons than any of the other major planets. With such a variety of planetary bodies to observe from close quarters, not to mention Saturn's iconic ring system, the urge to send orbiters was very compelling. As the nearer of the two, Jupiter was the first to be targeted. The *Galileo* spacecraft was launched on its six-year journey to Jupiter from the space shuttle in 1989. It operated successfully between 1995 and 2003 and was deliberately crashed into Jupiter at the end of its useful life.

An orbiter for Saturn was scheduled to follow, and Titan was firmly in the sights of the Saturn mission planners. The *Voyager* experience generated an overwhelming incentive to design a mission to the Saturnian system capable of discovering what lay below Titan's haze. Both the National Aeronautics and Space Administration (NASA) and the European Space Agency (ESA) were involved from early on with the conception and development of the mission; the idea from the beginning was to send an orbiter carrying a Titan probe. In what turned out to be a highly successful international collaboration, NASA provided the orbiter and ESA built the probe. The orbiter was named in honor of Giovanni Domenico Cassini, the French-Italian astronomer who discovered four of Saturn's moons and the gap separating the two main rings. The probe was named after Christiaan Huygens, the Dutch astronomer who discovered Titan. *Cassini* would be equipped with radar and infrared imaging capabilities for penetrating the haze; the independent probe was to parachute through the haze and radio back via *Cassini* the data collected by its instruments and camera.

No mission as complex as *Cassini–Huygens* had ever before been undertaken at such an immense distance from Earth. Even when Earth and Saturn are at their closest, the gulf between them is around 1.3 billion km. By the time *Cassini* was launched in 1997, *Galileo* had been performing well at Jupiter for nearly two years, even though its main communications dish had failed to unfurl correctly. But Saturn is roughly twice as far away as Jupiter. Light and radio signals take over an hour to make the one-way trip between Saturn and Earth, and even getting to Saturn at all would be less than straightforward.

*Cassini*'s route was necessarily a convoluted one. The 5.5-ton spacecraft was launched by a powerful rocket but could not make the journey in a reasonable time without extra impetus. To get some additional kicks, the mission design relied on "gravity assist"—a maneuvering technique whereby spacecraft pick up speed from close encounters with planets. Before it could set out properly on the main leg of its journey to Saturn, *Cassini* made two loops around the inner solar system. To gather enough speed, it skimmed close to Venus on two separate passes and then swung by Earth. Some two years after it had been launched on October 15, 1997, *Cassini* was finally catapulted away from the vicinity of Earth and toward the outer solar system. It received a final boost at Jupiter, about the halfway point. After being maneuvered into orbit around Saturn in July 2004, it embarked on a long series of loops, carefully planned to allow scrutiny of the planet, rings, and moons by its eleven instruments. If all went well, it would keep going for at least four more years.

On December 25, 2004, *Huygens* parted company with *Cassini* and for twenty days followed an independent orbit that would bring it close to Titan. Then, on January 14, 2005, *Huygens* plunged into Titan's atmosphere. As it descended to the surface, it transmitted data for two hours and twenty-eight minutes and conducted operations for over three hours after landing, until its batteries were dead. Unfortunately, after one hour and twelve minutes, *Cassini* was below the horizon and could no longer relay the probe's data back to Earth; also, a technical glitch caused the loss of some information for one experiment (though it was largely recovered by radio telescopes observing from Earth). Otherwise, to the delight of the triumphant science teams who anxiously monitored its progress, the probe worked almost entirely according to plan.

Even before *Huygens* reached Titan, the *Cassini* orbiter had begun its own program of mapping and remote sensing that would take it on dozens of close encounters with Titan. All eleven of its instruments were to be used to collect data on Titan; the expected deluge of information began to arrive on cue in the second half of 2004. The time had come to test the many hypotheses and speculations surrounding what would be found on Titan.

In the following chapters, we tell the story of how *Cassini* and *Huygens* have finally begun to lift the veil of mystery surrounding Titan, beginning with advancements in our understanding of Titan that took place in the decade preceding *Cassini*'s arrival. Some predictions have proved gratifyingly accurate; others have turned out to be misconceived, however plausible they may have seemed initially. Though many questions can now be answered—even some that no one thought to ask—they have quickly been replaced by a torrent of new and deeper puzzles.

But before we get to that story, we should set the scene. In broad brush terms, when was Titan discovered, what kind of world is it, and where in our solar system does it fit in the scheme of things?

## DISCOVERY

Born in The Hague in the Netherlands, Christiaan Huygens (1629–95) discovered Titan on March 25, 1655. He announced the existence of Saturn's moon a year later, and then went on to famously develop the wave theory of light and to become one of the greatest scientists of the seventeenth century. Besides possessing outstanding abilities as a theorist and mathematician, his many talents also included a practical bent. With his brother Constantyn, he designed and constructed a machine that could produce telescope lenses of better optical quality than any other at the time. Using a telescope made with one of these home-produced high-quality lenses, Huygens identified Titan as a moon of Saturn. Titan was the first planetary satellite to be discovered since 1610, when Galileo had found the four large "Galilean" moons of Jupiter, later named Io, Europa, Ganymede, and Callisto.

Though he discovered Titan, Huygens did not call it anything other than "Luna Saturni." For nearly two hundred years, the world we now call Titan was anonymous. The relatively small number of then-known planetary satellites were referred to by numbers. By the middle of the nineteenth century, however, new discoveries of more moons had rendered ambiguous the existing numbering system (wherein satellites were numbered in order of distance from their primary), so Sir John Herschel proposed the idea of giving moons individual names. From about 1848 on, astronomers happily adopted the names from classical mythology, including Titan, that Herschel had suggested.

## ONE OF A FAMILY

Living up to its name, Titan truly dwarfs the rest of Saturn's natural satellites. In sheer size, Titan shares more in common with its four substantial cousins in orbit around Jupiter. What its siblings lack in size, however, they make up for in number. As we write, the total count of Saturn's moons is at least fifty-six. The number of known satellites began to rise dramatically in 2000 because the Saturnian system was under close scrutiny from Earth in advance of *Cassini*'s arrival. In 2004, *Cassini* itself took up the search and found yet more. Saturn, perhaps more than any other planetary body, prompts the question, What is a moon? After all, each of the countless millions of ring particles is a distinct, rigid body, following its own orbit around Saturn, but it would be silly to call them all satellites.

Titan's size was not determined conclusively until the flyby of *Voyager 1* in 1980. Early estimates were all based on the tricky business of measuring Titan's apparent diameter when it is seen as a "flat" disk in the sky. (This measurement is tricky because Titan is dark toward its edges, unlike the disk of the Moon, which is nearly uniformly bright right to its edges.) The best measurements indicated a size of about two-thirds of an arcsecond (about the size of a golf ball eight miles [13 km] away). The opaque atmosphere further complicated the issue by making the visible disk look larger than Titan's solid body really is. As a result, its diameter was overestimated. Experiments conducted during an occultation in the

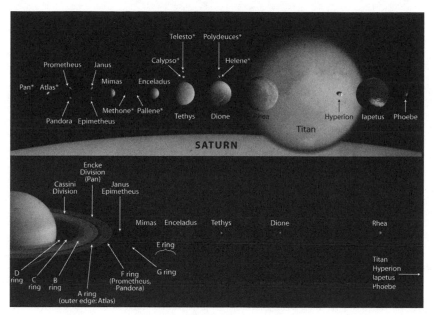

Figure 1.02.   Titan's family to scale. Titan completely dwarfs its fellow moons and lies well outside the rings and most of the moons. (NASA/JPL/Dave Seal)

1970s, when the Moon crossed in front of Titan, produced a figure of 5,800 km. For a time, Titan was thought to be the largest moon in the solar system, but then it was demoted to second place in the rankings when the results came back from *Voyager 1*'s radio science experiment.

As *Voyager 1* passed behind Titan's atmosphere, the spacecraft's radio signals were first deflected (though not blocked) by the moon's atmospheric gas. Analysis of the degree of deflection provided information on the temperature and pressure of Titan's atmosphere at various altitudes. Then, the spacecraft's radio signals were cut off completely when the spacecraft went behind Titan's solid globe. With these data, Titan's true diameter could be assessed: 5,150 km (to within 1 km), or 60 km less than that of Jupiter's moon Ganymede.

Titan's mass was first estimated in the nineteenth century by its effect on the orbit of Hyperion, the next Saturnian satellite out from Titan. The effect of Titan's gravity on the trajectory of *Voyager 1* allowed an even more precise measurement. Combining its size and mass ($1.346 \times 10^{23}$ kg) tells us that the average density of Titan is 1.88 times that of water, which is slightly higher than that of any of Saturn's other larger satellites. By comparison, the value for our rocky Moon is 3.34, and Earth, with its iron core, has a value of 5.52. Considering average density alone, Titan must be some mixture of ice and rock. Most likely, it consists of a rocky core overlain by a mantle chiefly made of ice.

It is no surprise to find that Titan, like all other satellites in the cold outer solar system (apart from Jupiter's exceptional moon, Io), has a substantial proportion of ice. The temperature at Titan's surface is around 94 K (or −179°C). Solar heating is so feeble and temperatures are generally so low that water ice is as hard as rock is on Earth—although like rock on Earth, the ice may be soft or even molten in the deep interior of Titan.

Jupiter's volcanically active moon Io (diameter 3,642 km), closest to Jupiter of its four large satellites, is the odd one out among the satellites of the outer planets, particularly with regard to composition. Io is made of rock and sulfur, and has virtually no water. Its interior temperature is raised to melting point by tidal energy resulting from its orbital motion within Jupiter's powerful gravity field. (The mechanism of tidal heating is similar to the way the gravity of the Sun and Moon raises tides in Earth's oceans.)

Europa (3,130 km), the second of Jupiter's Galilean moons, must be principally rock according to its average density. However, its surface layers are mainly water. Although its outer crust is frozen, a great deal of evidence strongly suggests that the crust floats on a global ocean of liquid water. Like Io, Europa is heated below its surface by tidal energy.

The other two large moons orbiting Jupiter, Ganymede (5,268 km) and Callisto (4,806 km), both have a higher proportion of ice and are more like Titan in this regard, though Ganymede's density is a bit higher than Titan's and Callisto's is a little lower. Unexpectedly, magnetic measurements made by the *Galileo* spacecraft hinted that Callisto, like Europa, may have an internal ocean, even though the tidal heating it experiences is not nearly as great as Europa's. These measurements also raise the intriguing possibility that Titan might have a subsurface ocean too.

Titan's more immediate neighbors in the Saturnian system each have individual characteristics and mysteries of their own, but looking at them alongside Titan only emphasizes the unique qualities of exceptional size and atmosphere that make Titan particularly fascinating. And if these lesser worlds have such varied and unexpected features, what greater surprises might be waiting on Titan?

The second largest Saturnian moon, Rhea, is only 30 percent the size of Titan (and only one-sixtieth the mass). It is one of six satellites in the

TABLE 1.1

Titan Compared with Other Large Moons, the Terrestrial planets, and Pluto

| | Average Distance from Sun (million km) | Year (yrs) | Radius (km) | Density (kg/m3) | Surface Gravity (m/s2) | Atmospheric Pressure (mbar) | Atmosphere | Escape Velocity (km/s) |
|---|---|---|---|---|---|---|---|---|
| Titan | 1,429 | 29.5 | 2,575 | 1,880 | 1.35 | 1,496 | | 2.64 |
| Mercury | 57 | 0.16 | 2,440 | 5,400 | 3.71 | – | | 4.24 |
| Venus | 108 | 0.61 | 6,051 | 5,250 | 8.96 | 90,000 | $CO_2$ | 10.37 |
| Earth | 149 | 1.00 | 6,379 | 5,450 | 9.80 | 1,013 | N2, O2 | 11.13 |
| Mars | 227 | 1.88 | 3,397 | 3,940 | 3.77 | 6 | $CO_2$ | 5.04 |
| Pluto | 5,913 | 249 | 1,137 | 2,050 | 0.66 | 0.03 | $CH_4$, $N_2$ | 1.22 |
| Moon | 149 | 1 | 1,737 | 3,340 | 1.64 | – | | 2.37 |
| Io | 778 | 11.87 | 1,815 | 3,550 | 1.82 | – | | 2.56 |
| Europa | 778 | 11.87 | 1,569 | 3,010 | 1.33 | – | | 2.04 |
| Ganymede | 778 | 11.87 | 2,631 | 1,940 | 1.44 | – | | 2.74 |
| Callisto | 778 | 11.87 | 2,403 | 1,860 | 1.26 | – | | 2.45 |
| Triton | 4,504 | 165 | 1,350 | 2,070 | 0.79 | 0.02 | $N_2$ | 1.45 |

| | | Orbital Distance (thousand km) | Orbital Period (days) | Inclination | Eccentricity | Mass (kg) | Radius (km) | Density (kg/m3) | Albedo |
|---|---|---|---|---|---|---|---|---|---|
| Mimas | (SI) | 185 | 0.94 | 1.53 | 0.0200 | 3.75E + 19 | 199 | 1,140 | 0.5 |
| Enceladus | (SII) | 238 | 1.37 | 0.02 | 0.0045 | 7.30E + 19 | 249 | 1,120 | 1 |
| Tethys | (SIII) | 295 | 1.88 | 1.09 | 0.0000 | 6.22E + 20 | 530 | 1,000 | 0.9 |
| Dione | (SIV) | 377 | 2.73 | 0.02 | 0.0022 | 1.05E + 21 | 560 | 1,440 | 0.7 |
| Rhea | (SV) | 527 | 4.51 | 0.35 | 0.0010 | 2.31E + 21 | 764 | 1,240 | 0.7 |
| Titan | (SVI) | 1,221 | 15.94 | 0.35 | 0.0292 | 1.35E + 23 | 2,575 | 1,881 | 0.21 |
| Hyperion | (SVII) | 1,481 | 21.27 | 0.43 | 0.1040 | 1.77E + 19 | 205 × 130 × 110 | 1,400 | 0.23 |
| Iapetus | (SVIII) | 3,561 | 79.3 | 7.52 | 0.0280 | 1.59E + 21 | 718 | 1,020 | 0.05–0.5 |
| Phoebe | (SIX) | 12,952 | 550.5 | 175.3 | 0.1630 | 4.00E + 18 | 110 | 700 | 0.06 |

400–1,500 km class, which are massive enough to have shapes close to spherical. In order of size they are Rhea, Iapetus, Dione, Tethys, Enceladus, and Mimas. Their predominantly icy nature is confirmed by their densities: all of them are only a little denser than water. Before *Cassini*, virtually everything known about their surfaces came from the encounters of *Voyagers 1* and *2*, but one of the most thrilling aspects of *Cassini*'s bounty from the early phase of its mission was the spectacularly detailed and comprehensive images of these distinctive worlds.

Moving inward from Titan's orbit, we first encounter Rhea and then Dione. Between them is a certain resemblance. Both are heavily cratered, and on each of them the leading hemisphere (the side facing the direction in which the satellite orbits) is markedly different from the other side (the trailing hemisphere). Rhea's leading side is brighter and more heavily cratered. A network of bright streaks crosses darker terrain on the other half. Although the impression from *Voyager*'s distant views was that the streaks might have been bright material sprayed out from the interior, the close-up views of Dione from *Cassini* revealed that they are the bright edges of cliffs where the crust has fractured.

Tethys is next. Its most striking feature is a huge impact crater called Odysseus. With a diameter of 400 km, this basin is nearly half the size of Tethys, though its once deep relief has sagged over time. The other distinctive feature on Tethys is a vast canyon called Ithaca Chasma. About 100 km wide, 3–5 km deep, and 2,000 km long, it stretches three-quarters of the way around the moon's circumference. A darker, less heavily cratered belt of terrain crossed by cracks is evidence that activity of some kind has altered part of the surface in the past.

We can only speculate about the activity that altered Tethys long ago, but to the delight of *Cassini* scientists, Enceladus proved to be active now, right in front of our eyes. Multiple jets of water vapor and ice particles are spewing from surface cracks, dubbed "tiger stripes," in the south polar region. *Cassini* even flew through the plume, which extends upward as much as 500 km. Some source of energy—tides perhaps, or radioactivity—is warming the material escaping through the cracks and is driving the geyserlike eruptions. Enceladus, along with Dione, Tethys, and Mimas, orbits within Saturn's tenuous outer ring, the E ring. It seems that Enceladus itself is the main source of the particles that make up that ring.

Mimas is the innermost of the larger moons. The dramatic 140-km crater Herschel, with its central peak, makes Mimas's crater-saturated face unmistakable and has earned it comparisons with the Death Star of the *Star Wars* movies. The gravitational action of Mimas was also responsible for clearing material from the Cassini division, which separates Saturn's A and B rings.

Moving out from Titan, the next moon we encounter, between Titan and Iapetus, is Hyperion. Though not one of the larger moons of Saturn, it is certainly one of the most curious. It is the largest known moon anywhere to have an irregular shape. When imaged by *Cassini*, it looked for all the world like a cosmic sponge. The many craters on Hyperion seem to have been deepened by a process called "thermal erosion." Solar radiation warms up dark colored dust deposited in the bottom of the craters; the resulting heat tends to melt the ice and deepen the craters. As well as looking like a sponge, Hyperion has another spongelike property—a great deal of empty space inside. A density of only 0.6 times that of water means it must essentially be an icy pile of rubble.

Iapetus, too, is intriguing and different. Though roughly spherical, parts of it are squashed in, and a strange ridge 20 km wide and 13 km high runs for 1,300 km around its midriff. But the most bizarre thing about Iapetus is the contrast between its leading and trailing sides. A huge dark reddish-brown area called Cassini Regio, which covers much of the leading side, reflects no more than 5 percent of the light falling on it, while the other side and the poles are ten times brighter. The explanation for Iapetus's duplicitous character remains disputed. One favored theory is that the dark region is coated with a layer of dust that came from one or more of Saturn's numerous outer moons.

Beyond the large inner satellites orbiting in stately order in the ring plane, at distances ten to twenty times farther from Saturn than Titan, we find a rabble of smaller moons (typically 5–40 km across). Their orbits are tilted to Saturn's equatorial plane by large angles, and a group of them are in retrograde (backward) orbits compared with the rest. This is seriously abnormal behavior for planetary satellites, and it suggests they were not born alongside Saturn in the same way as the inner satellites were. Saturn enlarged its family by adoption sometime after it had condensed out of the solar nebula and had already developed its own primordial satellite system. The evidence points to the errant moons as once having been wanderers through the outer solar system. Straying too close to Saturn's gravitational influence, they found themselves captured. And it seems they did not arrive randomly, since there are several distinct groups made up of members whose orbits share common features. These subsets of related moons are picturesquely known as the "Inuit group," the "Norse group," and the "Gallic group" and have been named individually after characters in the mythology of the respective culture. With so many moons to name, the committee tasked with the responsibility clearly decided it was time to tap into new resources or face the danger of exhausting the supply of names from Greek and Roman mythology!

Phoebe deserves special mention. It was discovered back in 1899, orbiting far out from Saturn in an exotic retrograde orbit. For a century it appeared to be a lone oddball, though we now know it belongs to the Norse group of Saturn's outer swarm of diminutive moons. However, Phoebe is still distinguished by size; it measures about 220 km across, which makes it an order of magnitude bigger than the other outer moons and explains why it was discovered so much earlier. In some ways, Phoebe's situation was much like Pluto's. Pluto was a puzzling misfit among the major planets for more than six decades after its discovery in 1930. Sixty-two years later, it was found to be just one of the larger members of the Kuiper Belt of icy bodies beyond Neptune. Interestingly, the parallel between Phoebe and Pluto does not stop there.

Smart mission planners arranged for *Cassini* to take a close-up look at Phoebe from a mere 2,068 km away as the spacecraft approached Saturn in June 2004. What they saw was a heavily cratered moon with a varied surface composed for the most part of water ice, but also laced with minerals and organic compounds. Phoebe had been regarded as a prime candidate for the mysterious source of dark material coating Iapetus, so it was a puzzle that *Cassini*'s data showed Phoebe's composition not to be a match for the dark part of Iapetus. Phoebe turned out not to be like the rocky asteroids in the belt between Mars and Jupiter, but is instead more akin to Kuiper Belt objects. Perhaps Saturn captured a miniature Kuiper Belt all of its own.

Not all of Saturn's small moons lie on the remote fringe. A collection of them are much closer to Saturn and are actually within the ring system. Some share orbits with each other or with larger siblings. Pan circulates in the Encke division and tiny Daphnis in the Keeler gap, both within the bright A ring. Prometheus and Pandora are "shepherds," herding the F-ring particles into a narrow ribbon. Complex interactions between the particles that make up the rings and the inner moons govern their orbits as they jostle by each other, responding to countless gravitational tugs.

These, then, are Titan's immediate family. Titan, of course, does not necessarily share any of their individual characteristics beyond being chiefly composed of ice and located in the same corner of space. However, as a group they set the context for the environment in which Titan formed and evolved.

## MOON IN MOTION

Titan revolves around Saturn some 1.22 million km from the center of its parent planet, a distance equivalent to about twenty Saturnian radii. It is considerably farther out than the ring system. For comparison, the easily visible A and B rings extend to about 2.3 radii from Saturn's center. Though more distant from Saturn than the rings, Titan's orbit is, like the rings, around Saturn's equator—or very nearly so; it is tilted by only one-third of a degree. Rather than being precisely circular, its orbit is slightly elliptical so that Titan's distance from Saturn varies by 71,000 km, or 6 percent.

Tied to its parent planet, Titan makes a circuit of the Sun each 29.458 years. And because Saturn's equator is tilted to its orbit by nearly twenty-seven degrees, Titan too is tilted to the same extent, relative to its path around the Sun. This significant tilt (or "obliquity") means that both Saturn and Titan experience marked seasons, much as Earth does with a tilt about three degrees less. When *Cassini* arrived in 2004, it was late summer in the southern hemispheres of Saturn and Titan; the solstice had been back in 2002. By the time the mission reaches its planned finish date in 2008, the equinox will be approaching—spring in the northern hemisphere and autumn in the south.

If a planet is tilted, it goes through a cycle of seasons each time it circles the Sun regardless of the shape of its orbit. Earth travels on an orbit that is not quite circular. Its distance from the Sun ranges between 147 and 152 million km, and our closest approach to the Sun occurs around the third of January each year. However, this change in distance causes only a few percent change in the amount of sunlight reaching points on the Earth's surface, an effect that is much smaller than that of the tilt.

Figure 1.03.   A *Voyager* image of Titan in blue light. No surface details are visible. There is a faint difference between the brightness of the two hemispheres, and a dark ring around the north pole. The black dots are reseau marks. Images taken with the vidicon camera technology used by early space missions such as *Voyager* could be distorted, and the reseau marks allowed the distortion to be corrected. (NASA)

The same is not true of Saturn and Titan. Saturn's distance from the Sun varies from 1,347 to 1,507 million km. Between the two extremes, the strength of the sunlight at Saturn changes by around 20 percent. When *Huygens* arrived at Titan in 2005, about two and one-half years had elapsed since Saturn was last at perihelion—its closest approach to the Sun. However, the effects that this variation of solar radiation have throughout Titan's "year" are likely to be relatively subtle compared with the changes in Titan's atmosphere induced by the march of the seasons.

One of the few conclusions that could be drawn from *Voyager 1*'s almost featureless images of Titan was that the northern haze cover looked darker than the southern hemisphere. Perhaps it was a seasonal effect? A decade later, Hubble Space Telescope images revealed that there had been a switch and the southern hemisphere had become the darker. Subsequent monitoring captured further changes to the haze, as if with the seasons it is blown back and forth between north and south.

## ATMOSPHERE

The very first suggestion that Titan might have an atmosphere dates back to the beginning of the twentieth century. José Comas Solà studied Titan visually using a 38-cm telescope at the Fabra Observatory in Barcelona, Spain. On August 13, 1907, he made a sketch of Titan, which was published the following year. He wrote, "[W]ith a clear image and using a magnification of 750, I observed Titan with very darkened edges" and concluded, "We may reasonably suppose that the darkening of the edges demonstrates the existence of a strongly absorbing atmosphere around Titan."

What Comas Solà claimed to have seen was certainly at the absolute limit of the capability of the human eye, if not beyond it. No one else ventured to claim they could see as much, but in 1925 the British astrophysicist Sir James Jeans used the kinetic theory of gases he had developed to demonstrate that it was theoretically possible for Titan to have an atmosphere.

Jeans pictured the molecules of a gas whizzing about, colliding with each other and anything in their way. The hotter and the more lightweight they are, the faster they move and vice versa. With nothing to hold them back—a container for example—the molecules soon fly apart and the gas disperses. A layer of gas around a planet or moon has gravity pulling it down, but with no lid on the top, any molecules traveling fast enough can escape from the gravity and shoot off into space. If a planet is going to hold on to an atmosphere for a length of time comparable with the age of the solar system, the gas molecules have to be cold enough and heavy enough to be confined in the gravity trap. Jeans calculated that, in the frigid conditions surrounding Saturn, Titan's gravity was strong enough to have kept hold of an atmosphere for as long as the solar system had been in existence. But there was a proviso. Lightweight molecules such as hydrogen and helium would have escaped long ago. The atmosphere, if it existed, would have to consist of heavier substances such as argon, neon, nitrogen, and methane.

The first proof that Titan indeed has an atmosphere came when Gerard Kuiper's spectra taken in 1943–44 showed the distinctive signature of methane. Methane was not the whole story, though. Even a small amount of methane shows up strongly in a spectrum. By contrast, the gas that forms the bulk of Titan's atmosphere does not announce its presence so readily in spectra. Data collected by *Voyager 1* in 1980 showed that Titan's atmosphere is overwhelmingly made of nitrogen, the gas that accounts for 80 percent of Earth's atmosphere, with methane amounting to no more than 5 percent. Further, atmospheric pressure at Titan's surface is 1.5 bar—that is, 50 percent greater than the pressure at Earth's surface. All at once, with regard to composition and pressure, Titan was understood to be the world in the solar system with the most Earth-like atmosphere.

Methane may be a minor constituent in Titan's atmosphere, but chemically, its importance is immense. A methane molecule consists of one carbon atom and four hydrogen atoms. In ultraviolet light, it breaks up into one or two hydrogen atoms and a fragment containing the carbon atom and the remaining hydrogens. The freed-up hydrogen ultimately tends to escape from Titan in the form of $H_2$ molecules, though not so easily from the much stronger gravity of Saturn. The hydrocarbon fragments combine with each other and with nitrogen in countless different ways to produce a vast range of organic compounds. Many were identified and their concentrations determined from infrared spectra recorded by *Voyager 1*. The most abundant are ethane ($C_2H_6$) and acetylene ($C_2H_2$). This process of photolysis is the origin of Titan's haze.

Haze particles form high in Titan's stratosphere where sunlight can penetrate to break up the methane. The concentration of organic molecules builds up to a level where clumps condense outward and stick to each other, then begin to drift downward very slowly. In fact, the haze is not really very dense. It is opaque chiefly because there is such a thick layer of it. *Huygens* found haze particles all the way down to the surface, although the atmosphere was clear enough to get good images of the ground from heights of 40 km and below. The logical consequence of the steady drizzle of haze is that Titan's surface becomes contaminated all over by a layer of organic sludge.

## METEOROLOGY

Here on Earth, the interplay between the atmosphere, the oceans, and radiation from the Sun makes for dynamic and often dramatic weather. Weather is so important to human well-being that monitoring it, predicting it, and just talking about it are major preoccupations. Since Titan has an atmosphere as impressive as ours, could it too have interesting weather?

Our weather is all about water and a chance combination of natural circumstances. It is a matter of everyday experience that water can be a liquid, a vapor, or a frozen solid in the range of temperatures encountered on Earth. What is less commonly remembered is that atmospheric pressure affects water's boiling point, though anyone who has used a house-hold pressure cooker or tried to brew tea at a high altitude will be familiar with the idea. So on Mars, for example, where the pressure is only 0.007 of that on Earth, water's boiling point is barely above 0°C. Liquid water put there would literally boil and freeze at the same time. Under Earth's atmospheric pressure the situation is entirely different, since all of 100°C separate water's freezing and boiling points. On Earth, of course, we have vast reservoirs of liquid water in the oceans, lakes, and rivers. Water evaporates from them, then condenses into clouds of droplets from which rain falls—or hail or snow if it is cold enough. This hydrological cycle combined with winds effectively moves water around our planet, including from oceans to land.

So how does Titan's weather compare? Unlike on Earth, liquid water and water vapor are not generally present; it is far too cold. On Titan, water in its rock-hard state plays the role that silicate rocks do on Earth. But in such frigid conditions, one substance can mimic the behavior of water on balmy Earth: methane. In the Titan regime, methane is able to evaporate, condense into clouds, and fall as droplets of rain. Going up through the atmosphere on Earth or on Titan, the temperature falls until it reaches a minimum. This altitude, called the "tropopause," is about 10 km high on Earth; on Titan it is some 40 km above its surface. Above that level, the temperature rises again through the stratosphere. It is below the tropopause, in the troposphere, that clouds form.

Despite the theoretical reasons for thinking that Titan should have clouds, first proof of their existence under much of the haze was not easy. The elusive cloud finally appeared in 1995 when Caitlin Griffith and Toby Owen were making infrared observations of Titan in the hope of discovering something about its surface composition. On September 4 they recorded wildly anomalous data amounting to twice as much radiation as they would have expected at a wavelength of three microns—more than could possibly have come from Titan's surface even if it were perfectly white. By careful analysis of subtleties in the spectrum, they recorded that they could account for the anomaly if clouds at a height of 10–15 km covered about 10 percent of Titan. It seems, however, that such large clouds may be rare.

More recently, clouds on Titan have been observed regularly both by *Cassini* and by one of the 10 m ground-based W. M. Keck telescopes. Curiously, these clouds have been restricted to a stormy patch over the south pole and a band around Titan at a latitude of 40° south. Within this band, individual elongated clouds stretch for over 1,000 km, and clouds seem to bunch up at two particular longitudes. The clouds rise up to a height of about 40 km and then dissipate, all in several hours.

The realization that methane, and one of the by-products of the atmospheric photochemistry—ethane—could both be liquid on Titan, and were present in large enough quantities, led to the idea that substantial lakes or oceans might exist. It did not seem unreasonable to take the parallel with water on Earth that far. So seriously was the possibility considered, the *Huygens* probe was designed so that it could float and test the depth to the solid floor below it, should it come down on a body of liquid. As was widely reported in the media at the time, *Huygens* landed not in liquid but on soft ground with the consistency of loose, wet sand. The chemical evidence suggested that liquid methane was making this surface material damp. Nothing from *Huygens* or from *Cassini*'s images and soundings in the very early part of its mission confirmed directly the presence of oceans, lakes, or actively flowing rivers. But even those earliest results contained strong hints from the landscape that liquid had been at work and had left its mark. We return to the unraveling of this Titan enigma in a later chapter.

## LANDSCAPE

*Huygens* and *Cassini* have transformed years of speculation about the nature of Titan's surface into a gradual discovery of the truth. *Cassini* is filling in detail on the tentative broad-brush maps begun in the mid-1990s with observations made by the Hubble Space Telescope, as well as ground-based infrared images. On a large scale, even the pre-*Cassini* observations revealed a highly varied surface with contrasting dark and bright areas. The largest and most conspicuous bright region is Xanadu, a plateau about the size of Australia just south of the equator. An extensive dark area, Shangri-La, lies to the west of Xanadu. (*Huygens* landed on the western periphery of Shangri-La.)

Features on Titan like Xanadu and Shangri-La, which are distinguished by their albedo (noticeably dark or bright areas), are named after sacred or enchanted places in world mythologies and literature. As happens with all planetary bodies, a system had to be devised for categorizing the types of features and giving them individual names. This is often a tortuous process because the informal names researchers adopt as soon as they spot interesting things on images have to be translated into "approved" names within the system. So, for example, the crater on Titan informally dubbed "Circus Maximus" became the Etruscan goddess Menrva, since the rules say that craters are to be named after "deities of wisdom."

The most detailed views of Titan's landscape came from the images recorded by *Huygens* as it descended, and from *Cassini*'s imaging radar, which at each pass cuts a swathe across the surface a few hundred kilometers wide and several thousand kilometers long. Even the earliest sweeps revealed a variety of features as diverse as we have on Earth, including windblown dunes, craters, chains of mountains, and what looks like a volcanic dome. Most intriguing of all are the channels, apparent shorelines, and lakelike features that say liquid has flowed on Titan in the past, or does flow from time to time even if it does not stand in oceans and large lakes at the present. *Huygens* landed close to what resembles a shoreline between bright and dark areas, and returned pictures of dark-colored branching channels cutting across the bright area.

## AN EARTH-LIKE WORLD?

The thrill of uncovering any extraterrestrial world and the challenge of demystifying the solar system piece by piece is enough to stimulate interplanetary exploration. But as humans, we cannot help being extra-fascinated by places that remind us of our home planet in some way. Mars is one such place. Titan, though so far away in the frigid depths of the outer solar system, is another. The argument that Titan can tell us something special about Earth and how life emerged here stirs the enthusiasm of many who find the solar system's inactive and unchanging balls of rock and ice less stimulating. Titan has been much vaunted as an Earth analogue in a variety of ways. But how far does the comparison go?

One thing planetary scientists agree on: Titan is not like an Earth in the deep freeze. The bulk compositions of the two bodies are totally different and reflect their origins at vastly different distances from the Sun. Nearly all would agree that Titan is not a likely home for extraterrestrial life despite the complex organic chemistry, though some have pointed out that niches for life might be created if internal or external heating causes water to melt.

Arguably, the most significant similarity between Earth and Titan is the likeness of their atmospheres. The presence of an atmosphere dramatically changes the nature of a planetary surface. There are physical processes, such as erosion and deposition of surface materials by wind and weather, and the temperature regime that would exist in the absence of any atmosphere is altered. And then there is the chemical interplay of the atmosphere with the substance of the solid surface beneath it and electromagnetic energy and high-speed particles bombarding it from above. Interactions of these kinds have shaped Titan as surely as they have characterized Earth. Of course, there is not an exact comparison between Titan's atmosphere and Earth's, since Titan's contains no oxygen and Earth's oxygen has come about through biological processes. However, an intriguing question is the extent to which Titan's present atmosphere resembles Earth's atmosphere in remote geological history. Certainly, the atmospheres both worlds possess now are not the same as ones they had in the remote past.

Today there is considerable concern about the buildup of so-called greenhouse gases, chiefly carbon dioxide, in Earth's atmosphere and the global warming that follows as a consequence. Titan has its own very potent greenhouse gas in the form of naturally occurring methane. There is also a sense in which Titan's haze plays a role similar to the ozone layer high in Earth's atmosphere, under threat from human disturbance of atmospheric chemistry. Though the chemical pathways involved are fundamentally different, the formation of both ozone and haze prevent ultraviolet radiation from penetrating nearer the surface.

But perhaps the most satisfying reward for all the endeavors to unveil Titan so far is the revelation of a world with landscapes bearing an uncanny resemblance to our home planet, despite the significant differences of temperature and composition. As the details are revealed one by one, it is sometimes hard to remember we are seeing ice in the role of rock and methane substituting for water. Titan really does provoke more comparisons with Earth than with any other planet or moon. In many ways we expected as much.

# 2. Waiting for *Cassini*

Up to the 1990s, the pace of discovery about Titan had largely been slow and patchy, the high points being the flybys of the *Voyager* spacecraft in 1980 and 1981, despite the disappointment of no surface features being detectable. But then, from the mid-1990s, the intensity of scientific activity directed toward Titan increased dramatically. This was not just in anticipation of *Cassini*'s arrival but was also a consequence of the impressive developments in telescopes and instruments of all types. As an intrinsically interesting and dynamic object, Titan was a natural target on which to focus with newfound capabilities and enhanced sharpness of vision. In parallel with observational developments, progress was made on the theoretical front as well. While *Cassini* was in transit across the solar system, ground-based astronomers were busy adding to their store of knowledge about Titan, all the better to plan *Cassini*'s observations and clarify the scientific questions to be addressed. In this chapter, we examine the state of knowledge about Titan at the time when *Cassini* and *Huygens* arrived, noting especially the developments that were then the most recent since the mid-1990s.

..............................................................................................

## RALPH'S LOG, DECEMBER 28, 2003

### Mt. Bigelow, Arizona

*Breaking and Entering*. Planetary science is a multidisciplinary endeavor, to be sure, but who would have thought that skills in breaking and entering have their use in exploring Titan?

The telescope is one of the biggest that mere mortals get to "drive" themselves, without the assistance of a full-time telescope operator. Officially the Gerard P. Kuiper 1.61 m telescope, it is known to everyone in Tucson—whether they think in metric or not—as "the sixty-one-inch." Observing here is probably closest to people's image of the astronomer, perched atop a lonely mountain. (More and more observing these days is actually done remotely, over the Internet, and the largest telescopes have full-time operators to assist the astronomer.)

It is the first of four nights of observing Titan. (The latter three nights would turn out cloudy.) I have high hopes of actually measuring something. A month or so before, I had hoped to record a stellar occultation but was clouded out. I am also keen to exorcise my sixty-one-inch demons. Five years earlier, while attempting to observe some of the Jovian moons that were to be observed by *Cassini* on its way to Saturn, my inept inexperience had resulted in the destruction of a $10,000 CCD (charge-coupled device) camera on this telescope.

This time I have my own equipment, so whatever I do, at least I can't break the astronomy department's CCD camera. But there is a problem. Somehow during the reconfiguration of the telescope after its previous user, last night's Lithuanian with a spectrometer, a vital component has been mislaid. There is a nice solid mounting plate at the bottom of the telescope, but for my instrument it is supposed to have a cylindrical two-inch-diameter focusing stage attached. Previously, this part has either been attached, or at least sitting around in the dome for us to attach, but it is nowhere to be seen. Amid all the high tech of sensitive semiconductor detectors, I am thwarted by a missing chunk of metal.

I file a trouble report on the observatory Web page, and leave a message on the operations answering machine. I begin to think all is lost, that the fitting is probably in some observatory workshop back in Tucson, locked up for the Christmas holiday closure. I warm up my dinner (some rather excellent rosemary lamb my wife, Zibi, had roasted at Christmas) in the control room microwave while I decide whether to abandon ship and head back down the mountain. But before calling it a night, with some trepidation, I call Bob Peterson, the University of Arizona (UA) observatory operations manager. On a Sunday evening between Christmas and New Year's Day, he understandably doesn't sound ecstatic to hear from me, but is most helpful. He suspects that the part is in the (locked) vacuum pump room, about ten feet from me. But I don't have a key—why should I need it? (Indeed, why should the room be locked at all?) He calls the observer at the Catalina Schmidt Telescope, about twenty yards from me, in case he happens to have a key. No luck.

"Well," he says, rather to my surprise, "you could try taking the hinges off the door." (The only alternative being to drive over to another telescope a few miles away, where apparently there is a key hidden somewhere, but it is already dark and freezing, and I've never been there anyway, so I am not enthusiastic about this option. Alone on a cold, dark mountain, even the most rational scientist can be confronted by fears of alien abduction or ax-wielding maniacs prowling the woods.) Stretching the phone cord as I talk and we explore what to do, I note that I can't see the hinge screws. They're on the inside of the door jamb. "No, no," he instructs, "punch out the hinge pins." I hang up and dash upstairs into the dome where there is a rack of tools. A screwdriver of a certain size suggests itself for the task, and a modest hammer. With a little work (the bottom hinge being awkwardly close to the floor), the pins are out. It takes an indelicate amount of brute force applied through two large flat screwdrivers to lever the door out. The woodwork will never quite look the same again, but out it comes. And lo, inside the room, there it is: the two-inch eyepiece/instrument holder.

Some quick work with a couple of bolts and I am ready to go. And all this just for a Titan spectrum and some photometry that in all probability will show that Titan looks like it usually does. But how else, unless we look, will we know that there isn't some huge storm on Titan like the one in 1995?

Figure 2.01. A modern amateur image of Saturn and Titan. Inexpensive modern Webcams and CCD cameras put scientifically useful tools in the hands of amateurs, and stacking many short-exposure images helps to compensate for the shimmering of the atmosphere. Titan is visible as a tiny disk. Image processed by Jason Hatton from a Webcam movie acquired with a 30 cm telescope by Bob Haberman. Used with permission.

## A NEW CLARITY OF VISION

One of the main difficulties standing in the way of observing Titan from Earth is the very small size of the disk Titan presents to us. Although some techniques of measuring light, such as spectroscopy or photometry, do not depend on resolving detail, seeing details in a picture is what most people want. If an object is below a certain angular size, it is essentially impossible to resolve any details on it. Two factors are in play here. In the case of very small telescopes, like those used by amateurs, the limiting factor is the diffraction of light as it passes through the telescope's aperture. Diffraction determines how big the image of a point of light appears. The smaller the telescope, the worse its intrinsic ability to resolve detail. By contrast, bigger telescopes tend to be compromised instead by the shimmering of the atmosphere, which blurs and distorts images. Astronomers refer to these atmospheric effects as the quality of "seeing." The "seeing" limit on a good, still night at a prime mountaintop observatory might be a little better than an arcsecond, which is roughly the diameter of Titan, or in fact the diffraction limit of a good amateur telescope with an aperture of around 0.25 m.

Before the 1990s, the fact that ground-based telescopes were sitting beneath our restless atmosphere meant that there were essentially no images showing nuances of light and shade on tiny Titan other than those captured by the spacecraft that had flown past it. Arguable exceptions are drawings made by Comas Solà (see chapter 1) and by the sharp-eyed French astronomer Adouin Dollfus. The fast image processing achieved by the human eye—brain combination can catch some fleeting instants of clarity that a photographic plate or CCD cannot. So, other than sketch-

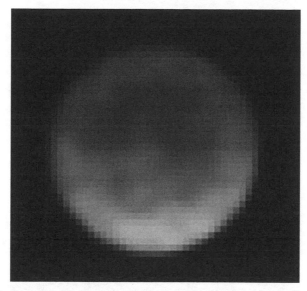

Figure 2.02. One of the first Hubble Space Telescope images to show surface details on Titan. It is about nineteen pixels across. This near-infrared (850–1,000 nm) image, taken in October 1994, shows the hazy south limb as particularly bright. Bright and dark patches are surface features.

ing, how do you reveal details on Titan and overcome the limitations imposed by seeing if your telescope is in principle big enough?

An obvious, if expensive, approach is to take the telescope above the interfering atmosphere. The Hubble Space Telescope (HST) took its first images of Titan in 1990. They were not outstanding, because HST's primary mirror suffered from a flaw that blurred its images (albeit in a precisely understood way), and exposures had to be short because HST had not by that time been programmed to track solar system targets moving against the stars. At least these pictures showed that Titan's atmosphere had undergone a seasonal change since the *Voyager* encounters.

In 1994, with its mirror defect corrected and a tracking ability installed, HST was used to make a set of images of Titan at a range of wavelengths. These showed clearly the seasonal change in the atmospheric haze—the northern hemisphere by then being the brighter one. But much more exciting was the fact that HST could also observe Titan at near-infrared wavelengths, allowing it to penetrate the haze and see down to the surface. By subtracting from the images the contribution from the atmosphere, which could be approximated reasonably well, a

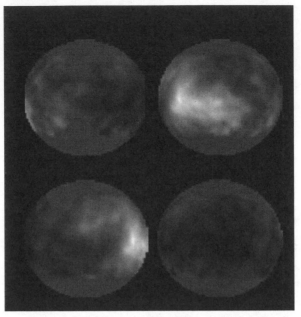

Figure 2.03. A map of Titan's bright and dark surface features made from fourteen HST images like that in figure 2.02. The map has been reprojected onto a sphere and is shown here as four views centered on points 90° apart. *Top left*, the sub-Saturnward point; *top right*, the leading face, showing the large, bright area Xanadu; *bottom left*, the anti-Saturn point (the *Huygens* probe landed near the center of this view); *bottom right*, the dark trailing hemisphere. Polar details cannot be seen. The images were taken near the northern autumn equinox when the equator was in the center of the image. (STScI/University of Arizona)

global map of the surface was extracted. The observations spanned two weeks: since Titan rotates synchronously—in other words, with the same side always facing Saturn—a day on Titan lasts one orbit period, or 15.945 days. Observing over a couple of weeks means that Titan makes nearly a complete revolution under the telescope, allowing a map showing the whole range of longitudes to be built up. This result produced a standing ovation at the 1994 DPS Meeting—the annual Division for Planetary Sciences meeting of the American Astronomical Society, which is one of the most important conferences in the business. Suddenly Titan was becoming a world with recognizable features, not just a small orange dot in the sky. Most prominent among the first surface features to be identified was a bright region the size of Australia on the leading face of Titan. This was the region now known as Xanadu.

An alternative approach to defeating atmospheric turbulence, and now the most popular, is the use of adaptive optics (AO). This is a technique

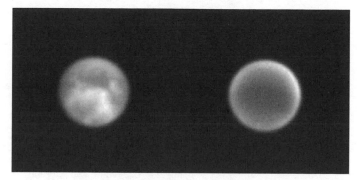

Figure 2.04. An example of the capabilities of modern (2004) ground-based adaptive optics, in this case the 8.2-m Very Large Telescope (VLT) operated in Chile by the European Southern Observatory (ESO). The image on the left is at 1.575 microns, in a near-infrared methane window that sees down to the surface; at right is 1.625 microns, deep in a methane absorption band so the only brightness is the high-altitude haze, in this case most visible at the northern limb.

by which the distortion created by the atmosphere is measured and compensated for in real time by a deformable mirror. It is as if the atmosphere were given an eye test and a custom pair of spectacles, fifty times a second. The technical wizardry behind this—the wavefront sensors, laser guide stars, control loops, and mirror actuators—are like bionic vision for a telescope. With the flick of a switch to engage the AO system, the dancing smudge of Titan suddenly sharpens into a crisp disk, showing details of its surface and atmosphere.

Astronomers discussing the performance of a telescope want detailed figures on the width of the point spread function (the size of the patch into which a point of light is smeared) and other parameters, but when it comes to explaining to politicians or journalists what is good or new about an observation or telescope, a picture of details on Titan (which is fairly bright and so easy to get quickly), with the sweeping, qualitative claim, "Before, this was possible only with the Hubble Space Telescope," has a simple persuasiveness to it. And so, Titan became the AO "poster child." As almost every AO system came on-line, the telescope would be trained on Titan to show off its capabilities.

The first AO images in the early 1990s were poorer than the HST ones, but the technology matured rapidly and HST was soon surpassed as far as imaging Titan was concerned, except at visible and ultraviolet (UV) wavelengths, where AO performs less well than in the near-infrared. HST, with its 2.4-m mirror, could achieve a resolution of about

Figure 2.05. Adaptive optics images of Titan from the 10-m Keck telescope at a wavelength where methane absorbs somewhat. These images do not probe all the way to the ground and show prominent clouds in the upper troposphere, particularly around the south pole. (H. Roe, Caltech)

0.1 arcsecond in the near-IR; a 10-m Keck telescope with its latest AO performance can do several times better. Coincidentally, the best AO images of Titan are about as good, in terms of number of resolution elements across the disk, as our naked-eye observations of our own Moon—good enough to see patches of bright and dark, but not quite good enough to figure out what they are.

It was always tempting to interpret the maps—perhaps Xanadu was some large volcanic highland, some of the dark patches might be large impact basins like those on the Moon. But there was no way to be sure without a closer look. Although the maps indeed got better and better between 1994 and 2004, our understanding of Titan's surface did not really improve. But the images told us much about the atmosphere, which displayed substantial changes over this period. Observing at different wavelengths allowed the haze amounts at different levels in the atmosphere to be measured, and the seasonal changes to be tracked. And the AO images in particular showed Titan's weather in action—clouds puffing up around the south pole.

Being able to resolve Titan's disk into more pixels is a prerequisite for distinguishing any detail there is to be seen, but as the *Voyager* spacecraft found, resolution on its own is not enough when faced by a world obscured to normal vision by opaque haze. HST and other telescopes, unlike the *Voyager*s, can now see details on Titan's surface, not just its opaque atmosphere, because they exploit particular chinks in Titan's hazy armor. Understanding how this works, and much of the progress in unpicking Titan's knotty problems, requires an introduction to Titan's spectrum.

## WINDOWS OF OPPORTUNITY

Visible light has a wavelength range of about 0.4 (blue) to about 0.7 (red) microns. There is nothing magical about this range—it just happens to be the part of the spectrum where the Sun radiates most intensely and the retinas in our eyes have evolved in response. The ultraviolet lies beyond the shortest visible wavelengths, and at wavelengths longer than visible red light the waveband is the infrared. *Voyager*'s cameras worked from about 0.35 to about 0.65 microns.

From space, we can see the surface of an airless body like the Moon—its patches of bright and dark, the shadows of mountains, and so on—at a wide range of wavelengths spanning the ultraviolet, visible, and infrared. However, gas molecules can scatter and absorb light, and do so both in our own atmosphere and in Titan's. Gas molecules scatter short (blue) wavelengths strongly. That is why our sky is blue. Titan has a thick atmosphere, mostly made of nitrogen like Earth's, so on its own it would look blue too. However, Titan's atmosphere is also laden with a thick organic haze, made by the action of solar ultraviolet rays on the trace of methane present, that slowly drizzles down and onto the surface. We can make similar haze in the laboratory by sparking or irradiating mixtures of methane and nitrogen. The stuff looks brownish, which means that the haze reflects red (and infrared) light, but absorbs blue and UV.

This explains the earliest space-borne observation of Titan—ultraviolet photometry from the Earth-orbiting satellite *OAO-2*. It showed in 1972 that Titan was in fact rather dark at UV wavelengths. It was already known from Kuiper's spectroscopy that it had a gaseous atmosphere, which should have made it bright in the UV. So something in the air, as it were, was preventing the sky from gleaming blue. This was the haze at work.

But the haze is much less absorbent of red and infrared wavelengths; the material is reddish in color after all. At longer wavelengths, the haze scatters rather than absorbs, so light can still fight its way through. To some extent, you can observe the same effect if you redecorate your house—it is easier to obscure markings on a surface with black paint, which absorbs all colors, than it is with white paint, which reflects all colors.

There is another effect too. The haze particles are 0.1 to 0.3 microns across, it seems, and such small particles are less effective at blocking radiation with a wavelength of 0.7 microns or longer. The haze's optical depth, roughly speaking the number of particles a photon of light will hit on average, drops from several at visible wavelengths to one or two in the near-infrared at 0.94 microns, and down to a fifth or so at 2 microns. There are in effect "fewer coats of paint" in the near-infrared.

So, because of the haze, Titan's atmosphere in broad terms is dark and absorbing at UV wavelengths, but brighter and clearer at red and infrared wavelengths. However, methane makes the picture more complicated. In the red and near-infrared regions there are narrow ranges of wavelength—bands—where methane absorbs. These are the dark bands observed by Kuiper that revealed the existence of the atmosphere in the first place. In one of these near-infrared bands, the haze reflects light quite well, but the gas absorbs it. In between the bands, in the so-called continuum or windows, the haze is bright and the gas is transparent.

Although the haze seems to be evenly distributed through much of the atmosphere, methane is concentrated in the lowest part. As a consequence of this, blue light and ultraviolet radiation are absorbed by the haze at high altitudes. Red light and continuum near-infrared are scattered a little by the haze but get down to the surface and can be reflected back up. Infrared in the wavelength range of a methane band, however, is scattered only a little by the haze but cannot reach the surface because it is absorbed by the methane. Understanding all this (and the quantitative details are still the subject of a lot of work) opens up some powerful possibilities. Since different methane bands, or different wavelengths in a given band, are absorbed to different extents, the light penetrates down to different levels in the atmosphere. In the deepest bands, the only light that is reflected by Titan is that which is scattered by the haze at the higher levels in the atmosphere, and so an image in one of these bands shows brightness where there is most haze. In a less deep band, one can see the high haze, and clouds in the lower atmosphere, but not the surface.

So what wavelengths are best for seeing the surface? The 1994 HST map was made at 0.94 microns, coincidentally the wavelength that cheap night-vision goggles or a typical TV remote control use. It was chosen because it is a continuum wavelength between two methane bands and is about the longest practicable wavelength for observing with the common detector in cameras, the charge-coupled device or CCD. At this wave-

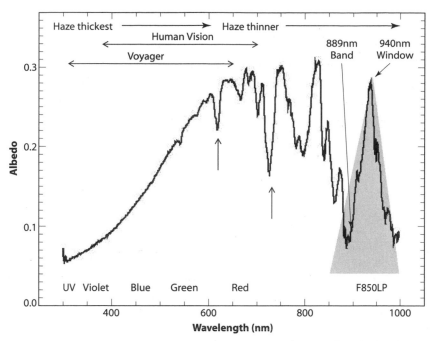

Figure 2.06. A schematic representation of Titan's reflection spectrum. The albedo is low in the UV and blue due to absorption by the dark haze; the haze is brighter at green, red, and near-infrared wavelengths. In the red and near-IR, progressively wider and deeper methane bands appear. The two arrows show the bands used by Kuiper to discover the atmosphere. The HST image and map in figures 2.01 and 2.02, and all the *Cassini* surface images in subsequent chapters, are taken through the 940-nm window.

length, the opacity of the haze is nevertheless substantial, and the perceived difference in brightness between bright and dark surface areas is only about 10 percent. One can just about see surface contrasts at shorter wavelengths, including the visible red, but the difference between bright and dark is only a couple of percent because the haze is thicker.

Now, with different detectors, such as those employed in specialized infrared cameras on large telescopes, or on *Cassini*'s VIMS (visual and infrared mapping spectrometer instrument; see chapter 3), one can exploit the other windows at longer wavelengths where the haze opacity is less. These windows are at 1.07, 1.28, 1.6, 2, 3, and 5 microns. The very small advantage over the 0.94 micron window offered by the one at 1.07 is not worth the added inconvenience of being unable to use a CCD. The 1.28-micron and 1.6-micron windows are rather better, and 2 microns turns out to be the best. The haze is even thinner at 3 and 5 microns than at 2, but there is less and less solar infrared at these longer wavelengths, so exposure times have to be longer. The 2-micron window turns out to be the best compromise. The first map at 2 microns was made by Roland Meier, working with Toby Owen and others in Hawaii using a near-infrared camera, NICMOS, on HST. The map showed impressive contrast, although it was soon surpassed in detail by the much larger telescopes being used for adaptive optics measurements.

.................................................................................................

### RALPH'S LOG, FALL 2001

Any book on Titan written before *Cassini* arrived was, of course, going to be out-of-date post-*Cassini*. That was never really a concern when we wrote *Lifting Titan's Veil*. It was intended largely as a primer on Titan, a report from the coal-face of science, and a historical record documenting what our thoughts on Titan were then and why we had them. But some things got out-of-date quicker than expected. One of them was the oft-quoted assertion that the haze hid Titan's surface entirely from the *Voyager* cameras.

In the late 1990s, several Titan workers began to wonder if this conclusion was not perhaps too hasty. As they worked with Hubble Space Telescope data and models of Titan's atmosphere, they began to realize that perhaps a small surface signal might be detectable in visible-light images of Titan. Eliot Young of Southwest Research Institute in Boulder, Colorado, and I were among a number of people suggesting that, should they ever find some spare time (ha!), it might be interesting to revisit the *Voyager* data.

It actually happened in 2001 in a graduate remote sensing class at the University of Arizona. Alfred McEwen, who is on the *Cassini* imaging team with responsibility for Titan surface observations, was teaching the class and giving students a small research project. One of the students, Jim Richardson, was assigned the project of trying to find Titan's surface in *Voyager* data. Alfred referred Jim to me for help, since I had worked on mapping Titan with HST and was familiar with

Titan. Our approach was essentially the same as had been used to make the maps of Titan with HST.

The question was, Had any light got through from the surface in the 600–640-nm wavelength band, the orange light to which the *Voyager* cameras were just sensitive? In a review paper I wrote with Jonathan Lunine in 1997, we noted that Chris McKay's radiative transfer code used to study Titan's atmospheric structure did indicate that Titan's observed albedo would be sensitive (albeit weakly) to the surface reflectivity at 640 nm, a point that had apparently gone unnoticed. Furthermore, Mark Lemmon's similar (but independently written) model also indicated such a sensitivity. Additional support came from the suggestion that a partial map made by Peter Smith and colleagues at 673 nm using Hubble Space Telescope images taken in 1994 showed the same broad pattern as that made at 940 nm, which showed much higher contrast. The 673 nm map was only a partial one, as only seven of the fourteen looks at Titan in that observation sequence included 673 nm images; the others were taken up with shorter-wavelength images to study the haze. Had there been more HST time, we might have made a complete 673 nm map, which may have led others to observe more closely at this wavelength too.

It looked like there was a chance. Jim waded through the *Voyager* image catalogs. The *Voyager*s took some two hundred images of Titan, although, of course, most appear similar and only a handful of different views are generally shown in papers and books. It turned out that the best images for our purposes were taken with the wide-angle camera, rather than the sharper narrow-angle camera, because the signal-to-noise ratio was better.

Jim used mathematical functions describing the variation of brightness from the center to the limb to boost the contrast of the images. Then, the contribution of a "synthetic atmosphere" of Titan had to be subtracted from the orange images that might have a surface signal buried in them. For our HST mapping, our synthetic average Titan was made just by averaging images from different longitudes. The *Voyager* images weren't suitable for this approach, however, so Jim used clear-filter and green-filter images as a background.

The subtraction left a noisy residual picture, with a noticeable north–south asymmetry. Nonetheless, a compelling pair of dark patches looked like they might be real. They were larger than the characteristic mottling of noise. Furthermore, the same patches appeared in three separate *Voyager* images, in the same places. They also seemed to coincide with some dark regions on the HST 673 nm and 940 nm maps.

In principle, this analysis could have been done when the *Voyager*s flew by, but one can understand why it was not. Computers then were not what they are now—and say you had found something; could you have believed it, and would anyone else have? Only subsequent weeks and years of Earth-based data showing that identifiable features rotate with Titan make a convincing case that they are indeed surface features.

In any event, the surface data extracted from the *Voyager* images added little to what was already known but was nevertheless an interesting exercise. Jim Richardson was what is referred to in Britain as a "mature" student. Ironically, given the imaging aspect of this project, he has a guide dog. In a previous career, as a nuclear reactor technician, he lost much of his sight in an industrial accident. His work is an inspiring example to the rest of us.

## TITAN'S ATMOSPHERE: THE BASICS

Before *Cassini*, most of our information about Titan's atmosphere came from *Voyager 1*. From observations made with Earth-based telescopes, it was known prior to that flyby that Titan has an atmosphere, that it contains methane, and that it is dark in the UV. But it took a close encounter with a spacecraft to learn what a substantial atmosphere it is, what its major constituent is, and how it works.

*Voyager*'s ultraviolet spectrometer found fluorescence characteristic of molecular nitrogen high in the atmosphere. Then, by measuring the refraction of *Voyager*'s radio signal to Earth caused by Titan's atmosphere, it was possible to show that, at Titan's surface, the atmosphere was about four times denser than the air at sea level on Earth. Additional information came from an infrared spectrometer on *Voyager*. Taken together, these data showed that the bulk of the atmosphere consists of nitrogen, as is the case for Earth. Titan's atmosphere was found to be more than 90 percent molecular nitrogen, much of the rest being methane. The surface temperature was measured as around 94 K (−179°C) and the surface pressure as about 1.5 bars. In terms of its vertical temperature structure, Titan looked remarkably like Earth. It was, to be sure, a very cold and vertically stretched version of the Earth, but all the basic features were there. Titan's atmosphere is far more extended because of the much lower gravity, which is only one-seventh as strong as Earth's.

From the surface upward, the temperature falls to a minimum value of about 70 K at the tropopause, some 40 km up. (On Earth, temperatures drop similarly with height, but to around −60°C or 213 K at a height of only 10 km above the surface.) This lower layer below the tropopause, the troposphere, holds most of the gas and contributes most of the greenhouse warming in the atmosphere. It is also where weather occurs—vertical motions and the formation of clouds. On Earth, most of the greenhouse effect is due to water vapor, and it is water vapor that condenses to form clouds of water droplets or ice, sometimes to fall down to the ground as rain, hail, or snow. On much colder Titan, the role that water takes on Earth is filled by methane.

Above the tropopause, the temperature increases. This makes the atmosphere stably stratified (hence the name, stratosphere) and little vertical motion occurs. The temperature rise is caused by the absorption of solar radiation. On Earth this is due primarily to ozone, which is formed by the action of solar ultraviolet rays on oxygen. On Titan, the same role is played by the haze. In fact, of the solar radiation falling on the top of Titan's atmosphere, only about one-third makes it down as far as the tropopause, and only some of that (about 10 percent of the total, or about one-thousandth the amount on Earth) gets down to the surface. So Titan was expected to be a very stagnant sort of place. Because its atmosphere is so thick, not only in density but also in height, it warms up or cools down only very slowly as the seasons change, and temperatures stay nearly constant over time.

Titan's equator is tilted relative to the path it takes around the Sun, along with Saturn, once every 29.5 Earth years. Titan is tilted only slightly more than Earth is, so it has seasons just like Earth's, only much longer.

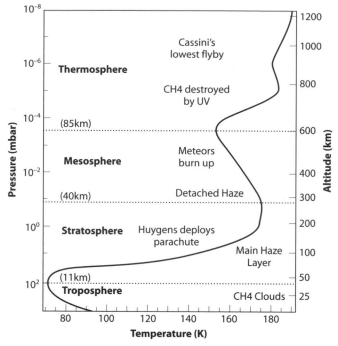

Figure 2.07. Titan's atmospheric structure, which resembles that of the Earth (albeit vertically stretched and much colder).

One of the most obvious manifestations of the seasons on Titan is a reversing wind pattern that pushes the haze around from one hemisphere to the other. As on Earth, solar heating is, on average, strongest at low latitudes but, during the peak of summer, instantaneously strongest at high latitudes. This heating causes warm air to rise, both on a small scale (producing the "thermals" used by glider pilots and soaring birds) and on a large scale, resulting in a generally cloudy band around the equator. The rising air has to be balanced by sinking elsewhere, and on the rapidly rotating Earth, this is at about 30° latitude in both hemispheres. This symmetric circulation pattern is called the Hadley cell. The descending air is dry, which is why we find our deserts at around 30° latitude.

Slowly rotating Titan's sluggish atmosphere behaves a little differently. In the stratosphere at least, where much of the Sun's radiation is absorbed, the flow rises over the summer hemisphere and descends over the winter one, In effect, there is a pole-to-pole Hadley cell. This situation seems to persist for almost half a Titan year. Then, at the equinox, as the subsolar latitude crosses the equator, the circulation must cease and re-

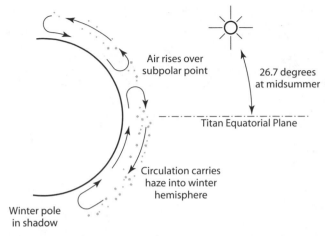

Figure 2.08. The meridional Hadley circulation during Titan summer. The warm air rises, per-haps causing clouds to form, and drags haze into the winter hemisphere. Unusual effects occur over the winter pole where low-temperature downward circulation brings organic com-pounds and haze down to lower altitudes.

verse direction. In so doing, models suggest, Titan may have a brief inter-lude, lasting an Earth year or two, when there is a vaguely symmetric Earth-like circulation, with rising at the equator and sinking in both hemispheres, before the reversed pole-to-pole motion takes over.

The summer-to-winter motion must drag some haze with it, and in-deed this is just what has been observed. Haze makes the atmosphere bright in the near-infrared, especially in the methane bands (where the haze is bright and above the absorbing lower atmosphere), and dark at blue-green wavelengths (where the gas would be bright but the haze is dark). So, a more hazy part of the atmosphere looks redder than a less hazy part—dark in blue light, bright in the near-IR.

When *Voyager* flew past Titan in 1980, the southern hemisphere was brighter to its cameras, which were primarily sensitive to short-wave-length light in the blue and green. When HST observed Titan in the 1990s, the situation was exactly reversed. The north was brighter. The appearance in the near-infrared (which *Voyager* could not see) was very different. First, instead of the limb (that is, the periphery of the visible disk) being darker, the limb was bright. It is easy to work out why. The haze is thinly distributed in a shell around Titan, so, viewing the edge of the disk, you look through a longer slanting path of haze than when you look straight down at the center of the disk. What you see is light scat-tered by the haze. More haze at the edge means more light. Hence, the

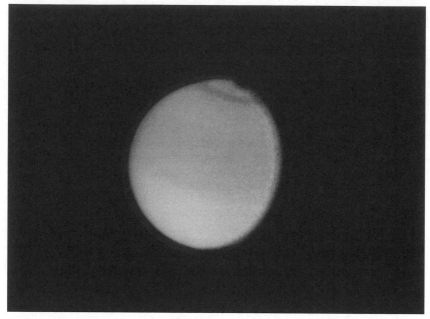

Figure 2.09. An image of Titan from *Voyager 2* (somewhat stretched in contrast compared with figure 1.03). The enhanced contrast brings out the polar collar, as well as the difference in brightness between the hemispheres. The detached haze is also weakly visible at the left limb. (NASA)

limb is bright in the methane bands, giving Titan a ringlike appearance. But it is not a symmetric ring. In the HST observations, more haze was in the south, and so the southern limb was brighter and thicker than the northern one. Titan appeared to have a smile.

As the models predicted though, the pole-to-pole Hadley circulation would drag haze back from south to north, and in the years around the turn of the millennium, the situation would reverse. In the twenty-first century, Titan quickly began to resemble once more how it had appeared to *Voyager* in the early 1980s, with the southern hemisphere bright in the blue. In the near-infrared, the smile became a frown.

The slow Hadley circulation of north–south winds is responsible for Titan's seasonal changes, but much faster zonal (east–west) winds are important too. These have the effect of shearing out clouds as they puff their way upward, and as we will see in chapter 6, they also manifest themselves in the transport of sand across the surface of Titan. They were important in the design of the *Huygens* probe's mission because the probe

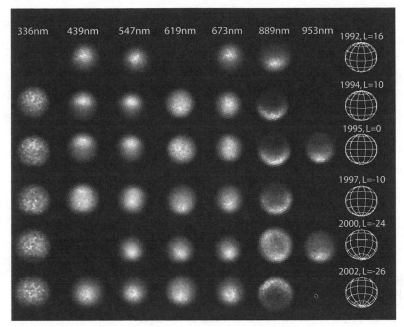

Figure 2.10. A collection of visible and near-visible images from the Hubble Space Telescope for the years 1992–2002. From left to right, they are near-UV (336 nm), blue, green, red methane band, red continuum, near-IR methane band (889 nm), and near-IR continuum (953 nm). The most prominent seasonal changes are in the blue sequence, where the brighter northern hemisphere progressively darkens relative to the southern one, and in the near-IR methane band images, where the "smile" becomes a frown as high-altitude haze migrates to the northern winter hemisphere. (R. Lorenz/STScI)

would drift several hundreds of kilometers eastward as it descended. These strong zonal winds occur to some extent on Earth (the jet stream that affects air travel across the Atlantic being an example), but they are most apparent in optically thick atmospheres, such as those of the giant planets or totally overcast Venus.

The first indications of the zonal winds were the deduction from *Voyager* infrared data that the equator-to-pole temperature gradient in the stratosphere required strong zonal winds for the pressure gradient to balance out. Although indirect, this model of Titan's zonal winds, developed by Mike Flasar of the Goddard Space Flight Center in Maryland, proved to be pretty close to the mark, suggesting winds of the order of 100 m/s in the stratosphere. Some additional support to the model came in 1989, when Titan occulted the star 28 Sagitarii (see later in this chapter).

So Titan, with its atmosphere, is in some ways a strange analogue of our home planet, the role of water being taken by methane. Titan is sometimes compared with the early Earth, before life and the appearance of oxygen. However, much warmer Earth was probably never as hydrogen-rich as is Titan, so this comparison shouldn't be taken too far. Nevertheless, the differences between the prebiotic Earth and the present Earth are certainly similar in nature to the differences between Titan and the present Earth, so there is doubtless much to be learned from processes on Titan that is relevant to understanding the evolution of Earth.

## TITAN'S ATMOSPHERIC CHEMICAL FACTORY

Despite the similarities, Titan's atmosphere is very different from Earth's with respect to chemistry. Even before *Cassini*, some twenty different organic compounds made of carbon, nitrogen, and hydrogen had been identified in Titan's atmosphere. These are nearly all made by the action of ultraviolet light on methane, aided and abetted, particularly for the nitrogen-bearing compounds, by cosmic rays and electrons trapped by Saturn's magnetic field. The simplest and most abundant ones include ethane, propane, and acetylene, all of which are present in natural gas on Earth. Also prominent is the poisonous gas hydrogen cyanide, at a level of about one part per million, not quite a toxic concentration for us. Many other more complicated molecules were detected from their infrared spectral signatures.

In contrast, oxygen-bearing molecules are conspicuous by their absence. The most abundant is the most volatile—namely, carbon monoxide—a gas also found in comets and probably around since Titan's formation. But less volatile oxygen-bearing species such as carbon dioxide and water are present only as traces—a few parts per billion. Even though these are very common in the warm inner solar system, at Titan they behave physically more like rock—hardly the stuff of an atmosphere. In fact, the traces that we do see in Titan's stratosphere probably did not come from Titan's surface, but from above, delivered as icy meteoroids that "burned up" as they lanced into the upper atmosphere. Some of the water thereby introduced is converted, again by solar ultraviolet-driven chemistry, into carbon dioxide. That we are able to detect the presence of these compounds at all is an impressive achievement. In abundance

TABLE 2.01
**Composition of Titan's Atmosphere**

| Common (official) Name | Formula | Amount (in stratosphere unless otherwise indicated) |
|---|---|---|
| Nitrogen | $N_2$ | 95% near surface <br> ~98% in stratosphere |
| Methane | $CH_4$ | 4.9% at surface near equator <br> 1.4% in lower stratosphere <br> ~2% at ~1000km |
| Hydrogen | $H_2$ | 0.1±0.2% in lower atmosphere <br> ~0.4% at ~1000km |
| Argon | $Ar^{40}$ | 43 ppm |
| | $Ar^{36}$ | 28 ppb |
| Ethane | $C_2H_6$ | ~20 ppm |
| Carbon Monoxide | CO | ~45 ppm |
| Acetylene (ethyne) | $C_2H_2$ | 3.3 ppm in stratosphere <br> 19 ppm at ~1000km |
| Propane | $C_3H_8$ | 700 ppb |
| Hydrogen Cyanide | HCN | 800 ppb in winter stratosphere <br> ~100 ppb in summer stratosphere |
| Ethylene (ethene) | $C_2H_4$ | 160 ppb |
| Carbon Dioxide | $CO_2$ | 15 ppb |
| Methyl Acetylene (propyne) | $C_3H_4$ | 10 ppb |
| Acetonitrile | $CH_3CN$ | a few ppb |
| Cyanoacetylene | $HC_3N$ | 5 ppb in winter stratosphere <br> <1 ppb in summer |
| Methyl acetylene | $CH_3C_2H$ | 5 ppb |
| Cyanogen | $C_2N_2$ | 5 ppb |
| Water vapor | $H_2O$ | 8 ppb |
| Diacetylene (buta-1,3-diyne) | $C_4H_2$ | 1.5 ppb (slightly higher in winter) |
| Benzene | $C_6H_6$ | 1.4 ppb at winter pole, <br> <0.5ppb elsewhere |

terms, they are present at the sort of level that man-made chlorofluorocarbons are present in Earth's atmosphere.

Tiny though the traces are, the amount of $CO_2$ in Titan's atmosphere appeared to demand rather more water to be delivered to Titan's atmosphere than was expected from models of how many meteoroids there should be. As we will see, this anomaly presaged later findings about the remarkable amount and mobility of water in the Saturnian system.

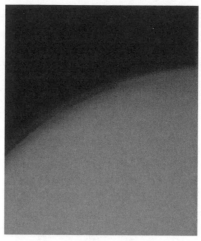

Figure 2.11. A close-up of Titan's northern hemisphere from *Voyager 1*. The polar hood appears to stand up above the main haze deck, and extends southward as a detached haze layer. (NASA)

Titan's atmospheric chemistry is not uniform and varies with both latitude and altitude. Since many molecules are produced by photochemistry high up, they are most abundant at high altitudes, and at latitudes where downward winds bring air from higher, more enriched levels. This happens most graphically during polar winter and is marked not only by a tenfold enhancement of the concentration of some gases but also by a dark region of concentrated haze. This haze seems distinct from that over the rest of Titan, having different spectral properties, perhaps because some compounds have condensed onto the haze particles during the polar winter. This phenomenon, called the dark polar hood, has some intriguing parallels on Earth. Here, the long winter night causes clouds of ice crystals to form, known as polar stratospheric clouds, and chemistry on the catalytic surfaces of the ice crystals results in local depletion of ozone. This ozone "hole" is isolated from the rest of the atmosphere by zonal winds, the so-called circumpolar vortex, which blocks flow to or from lower latitudes. Of course, the chemistry is very different on Titan (and indeed the dominant physical processes are too), but a detailed investigation of the polar hood is likely to be instructive for understanding the Earth too.

The polar hood was observed by *Voyager 1*, at the northern spring equinox, as a dark circle poleward of about 65° north. At the equinox, both polar regions were illuminated and visible, but there was no such feature in the south. From Earth, little could be seen with HST when

Titan was observed around the southern spring equinox in 1995. The HST images had a much lower resolution than the *Voyager* ones, and the polar region was right at the edge of the disk. However, as time went on, the south pole became more and more visible, and sure enough, a polar hood was apparent by the end of the 1990s. HST was equipped with a new camera in 2002, one able to produce better images in the ultraviolet than before, and it is at ultraviolet wavelengths that the hood shows strongest contrast. The hood was very prominent in the UV in 2002, but by 2003 it was beginning to fade. Of course, if Titan's appearance varies in the same way through each cycle of seasons, then the south polar hood would have to disappear by 2009 in order for Titan to look like it did in 1980. By the time *Cassini* arrived in 2004, it had in fact disappeared altogether.

## LOOKING FOR A NATURAL GAS SUPPLY

We've discussed how sunlight breaks down methane in Titan's atmosphere and turns it into ethane, acetylene, and other materials, notably the haze. But scientists interpreting the *Voyager* data realized that the entire inventory of methane in Titan's atmosphere would be destroyed this way in only about ten million years, or a fraction of the age of the solar system. So why was the methane still there?

Attention was drawn by the near-coincidence of Titan's surface temperature with the triple point of methane. At the triple point, materials can coexist in their solid, liquid, and vapor phases, and it so happens that Earth's average temperature is not far off the triple point of water, which is close to the freezing point of 273 K. The basic physics is straightforward, although there are some subtle and interesting feedbacks at work that have maintained Earth's temperature close to its present level for all of solar system history, even though the Sun was probably 30 percent fainter at the beginning. It takes energy (or "latent heat") to convert ice to water and water to steam. That is why ice cubes are good for cooling drinks, and why it takes a kettle a long time to boil dry. Much of the heat reaching Earth from the Sun goes into these phase changes, especially evaporating water, which drives the hydrological cycle. In cooler areas, the water condenses and falls as rain or snow, ultimately to be evaporated again and so on.

So, if Titan's temperature were close to the methane triple point, perhaps phase changes were buffering the temperature. By implication, phases other than the vapor one would be present. In other words, there might be liquid methane at the surface. The *Voyager* temperature and pressure data couldn't say for sure, but if there were seas of liquid methane on Titan, their slow evaporation could perhaps keep resupplying the atmosphere against the steady destruction by sunlight. Over 4.5 billion years, this process might use up around a kilometer depth of a global methane ocean. Such an amount—compare it with the 4 km average depth of the Earth's water oceans—was not implausible, but a 1-km-deep ocean would probably submerge most of Titan's topography. At most, there would be a few islands poking up before waves and tides wore them down.

It became clear in the 1990s, as near-infrared maps showed Titan to have bright and dark regions on its surface, and radar showed the surface was generally too radar-reflective to be covered in liquid hydrocarbons, that this global ocean model, appealing as it was, could no longer hold. Another idea put forward to solve the methane supply problem was that it is being continuously released from the interior, by volcanoes or geysers.

However, that didn't make Titan dry. The models of Titan's photochemistry, though they had all kinds of questionable assumptions in the detail, were, broadly speaking, pretty robust and predicted that most of the methane should be converted into ethane, which is also liquid at Titan's surface. Though ethane is produced only slowly by the atmospheric photochemistry, over the age of the solar system, several hundred meters' worth would be expected, along with a few hundred meters of solid stuff. The problem was not just finding some way of resupplying the atmospheric methane but also of disposing the ethane thereby produced.

Another inconvenient aspect of the no-ocean-volcanoes-instead model was that, unless the methane delivery rate was very finely tuned to match exactly the photochemical destruction rate, which varied through time as the Sun's brightness and color evolved, methane would accumulate anyway to the point where it would condense into lakes or seas, or it would be temporarily exhausted. The ideas of the *Voyager* era therefore faltered in the face of Titan's complex reality.

In space or time, somehow the picture is more complicated. Perhaps there is lots of liquid methane and ethane on Titan, but hidden in caverns and porespace beneath the visible surface, like the groundwater aquifers on Earth. Or perhaps, instead of drizzling down uniformly everywhere, the ethane has been dumped at Titan's high latitudes, which could not be seen from Earth, and maybe most of the methane is there too. Or maybe there hasn't been methane in Titan's atmosphere for all time— that would make for some interesting climate change through Titan's history, since the methane greenhouse effect is responsible for keeping Titan some 12 K warmer than it would otherwise be—in which case the ethane disposal problem is easier. Even with *Cassini*, this puzzle has not yet been solved.

## RADAR AND THE CASE FOR SEAS

A compelling piece of evidence in favor of seas on Titan came in 2002, when Titan moved into the sights of the giant Arecibo radio dish, operated by a group led by Don Campbell of Cornell University. Because this giant 300-m radio telescope is built into the ground, it can only observe over a narrow range of sky close to the overhead direction and so relies on planetary motion and the Earth's rotation to capture its targets. When Titan drew into view at Arecibo, the telescope had recently been upgraded with more powerful radar transmitters and sensitive receivers that would be able to do far better than Goldstone/Very Large Array (VLA) radar experiments carried out in the early 1990s.

These improvements meant that the signal-to-noise ratio was far higher. A measurement of the reflectivity of the disk taken as a whole would be more accurate and believable than the albedo determined before. And there was even enough signal to allow the echo to be chopped up in frequency and time.

The reason for separating different frequencies in the received signal is as follows. Titan spins with its rotation axis approximately orthogonal to the line joining Titan and Earth, so the morning edge of Titan is coming toward us while the evening side recedes. Thus, echoes from the morning side of Titan have a blue shift because the Doppler effect moves signal to a slightly higher frequency, whereas the opposite occurs on the evening limb. This allows us to measure how much of the echo is coming from the morning and evening parts of the disk, and how much from the center, which is moving neither toward nor away from us. (Of course, due to Earth's rotation, its motion around the Sun, Saturn's motion around the

Sun, and Titan's motion around Saturn, Titan's center is approaching or receding from us, but this is all taken into account separately.)

The radio telescope sent out a powerful microwave beam, at a single frequency. The beam's power was something like a megawatt. The Doppler spreading of the echo due to Titan's rotation caused most of the echo power to come back in a broad, hump-shaped spectrum with a bandwidth of about 375 Hz.

But strikingly, the Titan echoes had a unique feature. There was sometimes a strong spike in the Doppler spectrum, as if a very strong echo were coming back from just a narrow region in frequency space—in other words, from a small region on Titan. The only way this could realistically happen is if there were regions on Titan that are flat on the scale of the radar wavelength, so that there was a mirrorlike reflection.

So-called specular reflections like this are routinely seen by weather satellites on Earth's oceans, though only where the sea surface is calm. Although the intrinsic reflectivity of the ocean isn't in fact very high (which is why you can see a shallow seabed), the geometry of a specular reflection makes it appear very bright. And so it was with the Arecibo spike echoes. Although they were striking, the actual reflectivity calculated was quite low. These were dark mirrors.

A lake of liquid ethane seemed like an obvious, and appealing, explanation. And since photochemical models suggested that we should find lakes and seas of liquid hydrocarbons, the facts seemed to fit. But as always, there were other possibilities. A flat, organic-rich plain acting as a dark mirror could be all kinds of things, not just a lake of ethane with fluid properties much like gasoline at terrestrial temperatures. It might be a flat lakebed like the playas in the desert Southwest, but covered in a thick organic tar. Or it could be heterogeneous on scales smaller than the radar could see—small patches of bright, smooth reflection (flat ice, perhaps) dotted over a rough surface.

........................................................................................

## RALPH'S LOG, 2003

Because the radar observation of Titan was made possible by changing planetary geometry, with Titan slowly drifting into the sights of the Arecibo dish, it reminded me of a scene in the original *Star Wars* film (episode 4—"A New Hope"). The film's tense ending is paced by the inexorable motions of the Death Star and the jungle moon (Yavin 4), both in orbit around the giant planet Yavin. If the rebel heroes don't disable the Death Star in time, the moon, and the secret rebel base on it, will drift into the sights of the Death Star's super-laser (looking like a huge dish), and all will be lost.

While writing a commentary article for *Science* magazine to accompany the Campbell et al. paper, I cheekily threw in a reference to Yavin 4. This generated me some kudos among some of my planetary science colleagues, who, it must be admitted, are sci-fi fans. I am not aware of any other scientific publications mentioning this fictitious world. Some years later, findings on Titan would prompt me to write about another sci-fi word, "Arrakis," better known as "Dune."

## ANTICIPATING THE LANDSCAPE

Speculations about Titan's landscape had ranged from the exotic global ocean of hydrocarbons to a rather bleak and dull cratered iceball, like Callisto with an atmosphere. However, the variegated pattern of brightness at least showed that something was happening on Titan to make some areas light and some dark; but without any data to inform the speculations, they remained just that. It was possible to make some guesses—if so-and-so happened on Titan, then Titan's environment would be like this. One could, for example, anticipate how many impact craters Titan should have, by analogy with the crater populations seen elsewhere in the solar system. Titan would have very few small craters. Normally, small craters are the most abundant because the small comets and asteroids that make them are more common than big ones. But on Titan, small comets would break up in the atmosphere, as happened in the case of the Tunguska explosion in Siberia in 1908. Or, given the thick atmosphere and modest solubility of methane gas in water, one could predict that any volcanoes were more likely to be dome- or shield-shaped, rather than graceful cones like Mount Fuji: there wouldn't be enough gas to spray out "ash" fast enough to build a cone.

But it is not so much the character of individual processes like those just mentioned that would define what Titan's surface was like, but rather the balance between them. In the absence of any other information, one could at least estimate the amount of energy expended in different processes, considering the atmosphere and the planetary interior as engines that worked the landscape, driven by heat. Although it was difficult to relate a given type of feature to a given amount of work, at least the relative strengths of the processes could give some estimate of the landforms likely to dominate.

Consider the case of Earth, for example. About 80 mW m$^{-2}$ of heat leaks out of Earth's interior and is involved with the tectonic forces that push up mountains like the Himalayas and Alps, as well as more directly in the formation of volcanoes. A much higher amount of heat, about 20 W m$^{-2}$, flows across Earth's surface and drives ocean currents, winds, and waves. Maybe 1 W m$^{-2}$ is expressed as the tides in the ocean. And if the explosive energy in occasional impacts—one meteor crater of 1-km diameter every fifty thousand years, one dinosaur-killer every fifty million years, and so on—is added up, it amounts to only microwatts per square meter. Although each of these processes has a certain amount of energy associated with it, characterized by a numerical value, the efficiency of each in modifying the landscape is rather different. Impact cratering directly sculpts the surface, whereas the much larger atmospheric flows must find tools like gravel in rivers to make a mark on the landscape. But the progression of decreasing power, from atmospheric through tidal and volcanotectonic to cratering processes, is what defines our landscape. After all, impact craters occupy only a small fraction of Earth's surface, but river valleys and sand dunes occupy much.

The same calculation could be done for Titan. We know the amount of sunlight reaching the surface to drive the weather, and we know roughly the amount of rock and thus the amount of radioactive heat produced in the interior, which defines the volcanic/tectonic output. From these considerations, one finds the same progression: atmospheric—tidal—volcanotectonic—cratering, albeit with much smaller numerical values. Maybe the numerical values wouldn't matter so much, since Titan is made of more volatile, perhaps more easily worked, materials like ice and organics instead of rock. It was hard to say. Of most significance was the fact that the dynamic range of energies involved, the factor by which the atmosphere was stronger than the volcanotectonic work rate, and so on, was much smaller than that for Earth. In other words, Titan's competing processes were more closely balanced than on Earth, suggesting that we would see a varied landscape.

## A CLOUD OR TWO IN THE SKY

The improved telescopic observations of the 1990s shed light on one particular aspect of Titan's atmosphere. The detection of clouds indicated the existence of a methane "hydrological" cycle. It's worth emphasizing here an important difference between ethane and methane. Although both are gases at terrestrial temperatures, the boiling point of ethane is much higher than that of methane and it is essentially involatile at Titan's surface temperatures. Whereas methane can evaporate and form enough vapor to subsequently condense into clouds or rain, ethane just sits there. In principle, there might be ethane lakes—very old ones—but never ethane rivers. Any rivers would have to be created by methane rainfall.

The first firm observations of clouds, made in 1995 and 1998 by Caitlin Griffith and Toby Owen, relied on ground-based spectroscopy, which showed that Titan's brightness occasionally varied when measured at wavelengths at the edges of the methane bands. This was distinct from the regular, predictable variation in brightness observed in the windows between the bands due to the bright and dark regions on the surface coming into view as Titan rotated. These occasional variations could be dramatic over just a few hours. Further, by the way they differed between wavelengths, their location could be narrowed down to an altitude range between 10 and 30 km, just where methane would be expected to condense.

The fact that there were variations on timescales as short as a few hours suggested that these were convecting clouds, like the cauliflower cumulus clouds that puff up on a summer day on Earth. After the clouds blossom upward, rain forms in them and they dissipate. In fact, it turned out that HST had observed a large cloud at around 40° north in 1995, which was also broadly consistent with clouds being driven by direct solar heating. So, the changing clouds on Titan suggested an active weather cycle. However, it was known from *Voyager* that the lower atmosphere was not saturated with methane, so raindrops, which would fall slowly

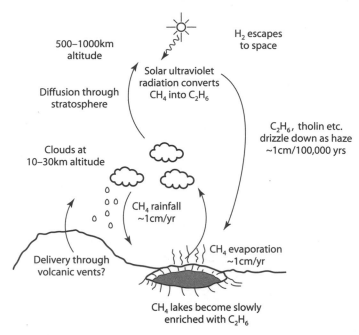

Figure 2.12. Titan's methane cycle. After its introduction into the surface—atmosphere system, perhaps through volcanic vents, methane participates in a slow hydrological cycle involving clouds, rain, and rivers. Some methane leaks up into the stratosphere, where it is converted into ethane liquid, which collects in lakes near the poles, and heavier solid organics, which drizzle down to the surface.

in Titan's low gravity, might evaporate before they hit the ground. Clouds didn't necessarily imply rain on the surface.

Titan's clouds turn out to be rather elusive and transitory. Part of the reason why they were hard to spot in the first place and why there have been few detections is that they aren't always there. That sounds trivial, but without continuous monitoring it is impossible to know and it is difficult to justify looking for something with an expensive and oversubscribed telescope unless you already have good evidence that you'll find it. So there was a chicken-and-egg problem over studying clouds if reliance was placed on the usual systems for allotting telescope time.

A keen observer might be able to get a run of a few nights on a telescope equipped to detect clouds on Titan. A generous allocation might be ten nights a year, but some of those could be hampered by bad weather on Earth or equipment problems. Let's say that Titan has visible clouds about half the time. If that were the case, there would still be a good chance of seeing clouds on at least a couple of nights. But clouds on Titan, it seems, are scarcer than that. Earth's cloud cover is about 30 percent, but some theoretical models of Titan's clouds and how often the weak sunlight can drive the upward convective air motions suggested a cloud cover of only a few percent. Multiply that by the number of nights of observing, and the odds of discovery are not so good.

The spectroscopic discovery of clouds at least showed they existed sometimes. But the fact that not all observers had seen them was a challenge and left us with no real idea of how big or how often methane storms were. It took some creative thinking to come up with a solution.

Astronomers, especially those studying distant galaxies or other faint objects, are used to long nights at the telescope, often staring at a tiny piece of sky to gather up every scrap of light. Or they are engaged in surveys, gathering statistics on large numbers of stars or asteroids. And so, telescope time is often allocated in blocks of whole or half nights. But with large telescopes, such as the Kecks, getting a good set of Titan images need take only fifteen minutes or so. But no formal way to get time in such short blocks exists.

In 2002, Antonin Bouchez at Caltech (and subsequently at the Keck Observatory itself), together with Mike Brown, came up with two clever approaches. First, Bouchez set up a command script at the Keck with all the right filter and adaptive optics settings, such that any regular observer could take a set of Titan images relatively easily if they happened to have some minutes to spare, perhaps before their intended target had risen high enough in the sky, or after they had got all the data they needed. Second, they found a way of getting at least a hint that there might be clouds worth seeing.

The early-warning system involved a fourteen-inch "amateur" telescope on the roof at Caltech and, more important, a diligent and talented graduate student, Emily Schaller. Although viewing conditions in the brightly lit and smoggy Los Angeles megopolis are hardly ideal, it was possible to take images of the Saturnian system with a CCD camera through two filters—one transmitting radiation inside a methane band and one in the nearby continuum. The murky skies make it a challenge to decide whether Titan is anomalously bright or not in an individual image, but assuming terrestrial murk affects both wavelengths the same, the ratio between Titan's brightness in the continuum and the band gives a much more robust measurement.

Even this was pushing the limit of what is reproducible. But after some months of observing, they had enough experience and statistics to be able to pull out Titan's underlying lightcurve and to see if Titan was anomalously bright. Armed with that information, a call to the Keck control room netted them a good number of cloud detections. Perhaps they were assisted by the fact that it was midsummer over Titan's exposed south pole in 2002–3, and clouds were bubbling up all the time.

This work stands as a remarkable example of what can be achieved with amateur equipment and some hard work. With the proof-of-concept (and proof in the results), this important program continues, now using a larger remotely controlled telescope in Arizona.

So ultimately, after the first, essentially unverifiable, reports of clouds on Titan in the 1990s, the large AO systems of the twenty-first century are routinely and unambiguously detecting clouds on Titan.


## EXPLOITING TITAN'S OCCULTATIONS

Occasionally, nature offers us an opportunity to secure high-resolution information about a moon or planet, even with a small telescope. For example, moons, planets, and asteroids are sometimes occulted by other solar system objects, especially the Earth's Moon. Jim Elliott of the Massachusetts Institute of Technology observed lunar occultations of Titan in 1974 and used his data to calculate a value for Titan's diameter.

Elliott's value of 5,800 km turned out to be an overestimate. Measuring Titan's size is complicated by the fact that Titan has an atmosphere. Not only is its edge fuzzy, its perceived diameter is not the same at all wavelengths. Since blue light is more effectively absorbed by the haze, the apparent diameter is larger in blue light than in red or near-infrared. The same wavelength effect was seen in images of Titan's shadow cast onto Saturn in 1995. At that time, Saturn's rings were presented edge-on, and when Titan was in front of Saturn, its shadow was projected onto the planet's disk. After taking into account that Titan's shadow was being cast onto the curved surface of Saturn's cloud deck, scientists were able to measure Titan's diameter at a range of wavelengths and to see that the distribution of haze with altitude wasn't the same all around the disk.

On occasion, Titan has itself been the occulting body, blocking out the light from a star for a short while. Precise timing of how long the star

Figure 2.13. An HST image of Titan (at left) and Saturn, during the ring-plane crossing season (southern spring equinox) in late 1995. The rings are seen edge-on, and several other moons are visible on the right. The unique aspect of this picture is the shadow of Titan cast onto Saturn. Careful study of the shape of the shadow revealed characteristics of the haze distribution that would otherwise be impossible to detect. (E. Karkoschka/STScI/University of Arizona)

winks out during such an occultation can define the diameter of an occulting body quite accurately, even if the body itself cannot actually be resolved with the telescope making the observation. Furthermore, if the occulting object has an atmosphere, as in the case of Titan, the way in which the starlight gradually declines at the start of the occultation, then rises again at the end, can be analyzed for information about the structure of the atmosphere.

In effect, the atmosphere behaves as a lens. As the beam of starlight travels through the atmosphere, its path is bent. Its intensity varies with time as it scans through different levels of the atmosphere allowing the density (or more strictly, the refractivity) to be measured. From this the temperature structure can be inferred. The technique only works over a specific range of altitudes—those where the air density is high enough to produce measurable bending, but not where the gas or haze absorbs the light to undetectable levels. Spikes in the lightcurve can also indicate local inhomogeneities in the atmosphere, such as gravity waves. On Titan, the altitude range probed by occultations is between about 250 and 450 km.

A major occultation event occurred on July 3, 1989, when Titan crossed in front of the star 28 Sagittarii. At magnitude 5.5, 28 Sag was bright enough to be seen by the naked eye under good observing conditions. Such an occultation is sufficiently rare and of such potential importance, considerable effort was put into the preparations to observe it. Larry H. Wasserman of the Lowell Observatory predicted that the occultation would be visible from Europe, but considerable uncertainties were inherent in the calculations. No one could be confident about what would actually be seen from any particular place. For once, astronomers found their luck was in. The prediction of when and where the occultation would be visible was pretty accurate and skies were clear. The event was recorded from as far north as Sweden to the Pic du Midi Observatory in the Pyrenees Mountains of southern France.

The intensity of starlight declined over about twenty seconds as anticipated, complete with "spikes," but then there was a total surprise. The occultation was expected to last some five minutes. About halfway through, there was a bright flash as if the star were shining through Titan. In fact, as the star got close to being directly behind the center of Titan, its light was being bent around the entire limb of Titan. The flash lasted around five seconds at Paris and a shorter time elsewhere. The shape and timing of the central flash provided a bonus of additional information about the atmosphere.

If Titan's atmosphere were perfectly uniform, a ring of images would be formed. As shown by a careful analysis of the data from this event, Titan's atmosphere seemed to act as if deformed by rapid rotation into an oblate spheroid, allowing in principle four simultaneous images of the star. The oblateness of the 250-microbar contour in Titan's atmosphere revealed by the occultation confirmed that the upper atmosphere must be rotating fairly fast, with zonal winds of perhaps 100 m/s at high altitudes, causing the atmosphere to deform. This was a valuable piece of information, confirming the indirect suggestion from the *Voyager* temperature measurements that Titan should have strong zonal winds. These meant that the *Huygens* probe could expect to drift several hundreds of kilometers during its descent.

Although occultations by Titan of a star as bright as 28 Sag would be expected once every fifty thousand years, seeing a central flash would be expected only once every million years! This was indeed a remarkably fortuitous coincidence. But more opportunities for stellar occultation observations followed, even if they were less spectacular.

The occultation of a much dimmer star took place in 1995, and an analysis showed Titan to be slightly off-center optically (compared with 1989). This suggested that the structure of the haze in relation to latitude was changing, just as the HST images were showing.

A rather favorable occultation occurred on December 20, 2001, when a 12.4-magnitude star was occulted by Titan high over the United States. Although this star was too faint for most amateur telescopes, an arsenal

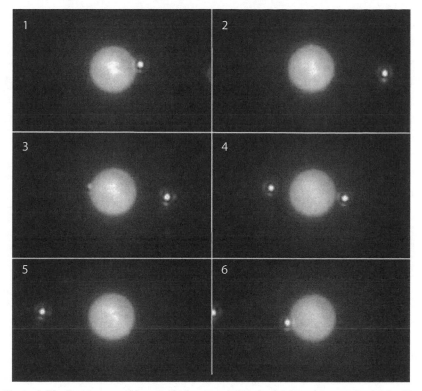

Figure 2.14. A sequence of images from a near-infrared adaptive optics movie acquired by Antonin Bouchez and colleagues at the Palomar two-hundred-inch telescope in late 2001. Titan's disk is clearly resolved, and the Xanadu bright feature is seen at the center. Titan was observed moving across a pair of stars (although all these images are centered on Titan). In some of the images, the stars are visible as bumps on the edge of the disk. Their precise position and brightness when close to the disk gave a picture of the structure of Titan's atmosphere. (Antonin Bouchez/Caltech)

of large telescopes was aimed at Titan, in particular the two-hundred-inch Palomar telescope with its new adaptive optics system.

The team there, led by graduate student Antonin Bouchez from Caltech, was lucky. The weather and the equipment both cooperated. Titan was seen as a disk, with Xanadu visible. The star was an unremarkable object, looked at in any detail only because Titan happened to pass in front of it (an unlikely claim to stellar fame); but it turned out to be two stars! They were separated by only a little over an arcsecond, so the regular telescopes that had observed it previously had not recognized it as a close pair of distinct stars. Bouchez feared for a while that there might be no occultation at all, that Titan might pass between the two stars.

Luckily, the alignment was almost perfect, and Titan moved across the two stars in turn. The first winked briefly as Titan's atmosphere passed in front, and its image could then be seen moving along the northern edge of Titan, before reemerging on the other side. The second star hit a little farther south, and its image passed along the southern limb. At some moments, the symmetry of the atmosphere allowed two images of the star to show, on opposite sides. Although neither star was close enough to dead center to cause a central flash, having the primary images of the stars in opposite hemispheres gave a global sampling of the atmosphere.

Less cinematographic, but nonetheless important, was a pair of occultations on November 14, 2003. These were the last chance to observe Titan this way before *Cassini* arrived, and mission planners were concerned about gravity waves in Titan's atmosphere. These periodic fluctuations in density might, if large enough and caught at an unlucky moment, cause the *Huygens* probe to trigger its parachute too early, possibly outside the regime where it would survive. Since gravity waves could be detected by the scintillations in an occultation, confirming that these fluctuations were not too large would give a useful boost to confidence in the mission.

The first of the occultations was of a fairly bright star of magnitude 8.6, within the reach of amateur telescopes. However, it was observable only in the more remote Southern Hemisphere. Observers, led by Bruno Sicardy of the Paris Observatory, set up a "fence" of telescopes in Namibia, South Africa, and Madagascar, with the hope of capturing the all-important central flash. Several telescopes caught it, with some adjacent telescopes making lightcurves at different wavelengths.

Later that night, when Titan could be observed over North America, a less generous 10.4-magnitude star was occulted. However, bad weather thwarted most observers (including the first author of this book, surrounded by clouds on Mt. Bigelow).

## TAKING TITAN'S X-RAY

On January 5, 2003, Titan executed a truly bizarre occultation when it drifted in front of one of the most exotic and energetic objects in the

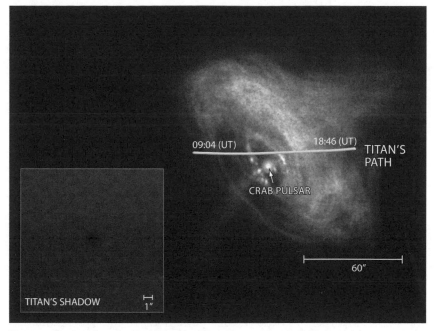

Figure 2.15. An image by the Chandra X-ray Observatory of the Crab Nebula, which is a bright, extended X-ray source. On January 5, 2003, Titan crossed in front of the nebula, and processing of the data allowed the one-arcsecond-diameter X-ray shadow cast by the moon (*inset*) to be reconstructed. (NASA/CXC/Penn State/K. Mori et al.)

sky—the Crab Nebula. This object is a supernova remnant, which was formed in the year a.d. 1054 by the explosion of a star that was witnessed and recorded by Chinese and Arab astronomers. The remains of the star itself are a pulsar at the heart of the nebula. Powered by high-speed electrons emitted by the pulsar, the nebula is one of the strongest radio sources in the sky and also a source of X-rays. The alignment of Titan and the Crab in 2003 was the first since the nebula was formed. The next will not occur until 2267. Fortunately, the once-in-a-millennium event was observed by the Chandra X-ray Observatory, a 20-m-long five-ton spacecraft launched in 1999 from the space shuttle into an orbit 10,000 km above Earth.

Koji Mori of Pennsylvania State University and his colleagues observed the Crab with Chandra for nine hours as Titan swept across. Even though the Crab is a powerful source, there are so few X-ray photons that one can't just see the shadow of Titan. Mori and his team reprojected the position of each photon onto a reference frame attached to Titan, in effect unsmearing Titan's motion across the nebula. After that, they had to consider how much the telescope itself smeared out the shadow.

They found that the X-ray shadow was about 0.588 arcseconds across, implying that the altitude above Titan's surface below which the X-rays were blocked was about 880 km. This is much higher than the altitude of 200 km below which visible light is blocked. The explanation is that X-rays, although better than light at penetrating short distances through solids, are less effective at penetrating long distances through air.

When scientists compared this atmospheric thickness with what would be expected from the models based on *Voyager* data, it seemed that Titan's atmosphere was a little thicker. Perhaps it had warmed and puffed up a little because Titan was closer to the Sun.

The planetary science community working on Titan had, of course, never crossed paths with these X-ray astronomers before. It was an ingenious observation, indeed, but no one had a good sense of how much to believe the results. Was the observation or its interpretation suspect, or had the atmosphere really puffed up? Perhaps it wouldn't be a surprise to learn that it had, since Saturn's orbit around the Sun is appreciably eccentric and Titan was closer to the Sun than it was when *Voyager* flew by.

### INSIDE TITAN

Although the details of Titan's surface remained elusive, at least some good theories were put forth about Titan's overall composition and the structure of its interior. The way we think planets and moons form is by collision of much smaller chunks of matter called planetesimals. These in turn are made of rock and ice, ultimately themselves made of tiny particles that stuck together in the swirling cloud of dust and gas around the Sun when the solar system formed. The proportion of rock, metal, ice, and frozen gases in these planetesimals presumably varied throughout the solar system, with the inner solar system dominated by rock and metal so as to give rise to the dense terrestrial planets. But further than about 4 AU from the Sun, beyond the so-called snow line, water ice becomes abundant, so most of the satellites in the outer solar system are icy. And water ice, if cold enough, can also trap gases such as ammonia and methane.

Each of the giant planets' retinues of satellites is a solar system in miniature. As the planetesimals (with whatever composition is character-istic of that distance from the Sun) float along in their orbits around the nascent planet, they begin to clump together, sticking like snowballs. But as the clumps become bigger, their gravity draws other planetesimals in at faster speeds, and soon they begin to smash together. As a satellite grows, the collision energy increases, to the point where incoming plane-tesimals are heated by their impacts such that the ice melts. And so, the growing Titan began as a core of gas-laden ice and rock, but was then surrounded by a layer of rock, where the rocky material in the later planetesimals had fallen through the molten ice. This, in turn, was over-lain by a layer of liquid water and capped with a primitive atmosphere. That initial hot atmosphere probably had little in common with the atmo-sphere we see today, being richer in water vapor and ammonia.

It sounds like a rather fancy story, supported by little more than the observation that many outer solar system satellites are icy. But the pattern seen in the solar system is repeated in microcosm around each giant planet, suggesting the formation process is rather consistent with the pic-ture above.

Gravity measurements by the *Galileo* spacecraft in the late 1990s began to show the overall structure of Jupiter's Galilean moons. Superimposed on the issue of bulk composition (Io being all rock, ice, and sulphur; Europa being like an Io with a 100-km veneer of liquid water and ice; and Ganymede and Callisto being more like half rock and half ice) is the question of how much their interiors have sorted themselves out. Even though the bulk ice fraction increases as one moves away from Jupiter, the ice fraction in the exposed surface (roughly indicated by the optical or radar reflectivity) decreases. Europa is so bright presumably because its surface is covered in fresh frost, whereas Callisto is dead and dark.

The question naturally arises as to why Titan has an atmosphere when Ganymede and Callisto do not. A number of factors are at work, and it is not really known which are the most important. First, the composition of the warm Jovian nebula was probably less volatile-rich, the cloud being too warm to retain much ammonia or methane in the ice. Second, the Galilean satellites are deep in the solar and Jovian gravity wells, such that planetesimal collisions would have been more energetic—enough not only to melt the ice but also to blow off nascent atmospheres. And if they ever formed in the first place, Galilean moon atmospheres would have been more susceptible to thermal escape or to erosion by Jupiter's radiation belts.

Some surprises and subtleties remain, one being the curious structure of Callisto, which appears not to have separated into a large rocky core and outer ice mantle, but is largely undifferentiated. And yet it seems to show evidence of having a liquid water layer beneath its dirty ice surface. As *Cassini* reveals more details about the structure and composition of the Saturnian satellites, perhaps a more robust picture will emerge about how planets and satellites form in general. Titan being the same size as Ganymede and Callisto makes it a particularly important piece of the puzzle. But the story of the solar system seems to be as much about exceptions as it is about rules.

Before *Cassini* arrived, the expectation was that Titan's structure might involve a solid crust of 50–100 km of ice, overlying a liquid water layer several hundreds of kilometers thick, much like the icy Galileans, but perhaps with some ammonia dissolved in it, acting as an antifreeze. Below that would be a layer of more ice, with a particular high-pressure crystal structure due to the weight of the ice and ocean above. Finally, beneath that would be the rock-metal interior.

But could the interior be hot enough to allow the metal and rock to segregate in the deep interior, to form a metal core like Io's? And how deep below the ice was the water-ammonia ocean? These would depend on details of convection, the thermally driven churning of a planet's soft interior, that continue to challenge modelers. And then, unlike the Galileans, the possible abundant presence of ammonia and methane could change the properties of ice to the extent that ices could float or sink in Titan's interior, depending on the ammonia content.

Methane and water ice have a particularly complicated relationship. Water molecules can form a frozen cage around methane and other gases, trapping them in a cagelike structure called a clathrate. Clathrates can be found in deposits on the Earth's cold ocean floor, from which the methane is occasionally belched out. Indeed, if a sample of this ice is brought to the surface and thrown into warm water, the methane will fizz out of the ice; it is even possible to set fire to it. Exactly how this material has behaved, and continues to behave, may be one of the most important factors in the history of Titan's atmosphere. For one thing, the initial core of cold ice-rock planetesimals might have retained methane while the rest of Titan accreted around it, and only hundreds of millions of years later did the ice warm and soften from the radioactive heat in the rock to let the gas out, to bubble through the water layer and perhaps to ultimately work its way up to the surface.

In 2001, John Loveday at the University of Edinburgh in Scotland and his colleagues reported new measurements on methane clathrates at high pressure. Such high pressures are not relevant for clathrates on the Earth's seabed, and it used to be thought that hundreds of kilometers down in Titan's interior, these pressures would cause methane clathrates to decompose, releasing the methane early in Titan's history. Using an intense synchrotron X-ray source, they studied clathrates held in a high-pressure diamond anvil cell, and found that methane clathrate had a new structure at high pressure and was in fact quite stable. This, in turn, meant that perhaps Titan's interior could hold onto its methane, allowing it to dribble out slowly in volcanoes, perhaps at the present day.

..........................................................................................

## RALPH'S LOG, SEPTEMBER–NOVEMBER 2003

MONTEREY, CALIFORNIA, SEPTEMBER 2003

In 2003 it is the turn of the pleasant seaside town of Monterey to host the DPS conference. The DPS (the Division for Planetary Sciences of the American Astronomical Society) has usually been the foremost annual conference for Titan news, at least on this side of the Atlantic. The other regular conferences, the Lunar and Planetary Science Conference (LPSC, always in swampy Houston) and the American Geophysical Union (AGU, with a small spring meeting on the East Coast and the giant fall meeting a couple of weeks before Christmas in San Francisco), have tended to have only one or two Titan talks. As of the following year, once the deluge of data begins, that would change. LPSC is basically a lithophile conference—geologists, meteoriticists—people who so far don't have much to say about Titan yet. Atmospheres and the solar system beyond Jupiter tend to get short shrift. But *Cassini*'s data will propel Titan into the purview of geologists. Titan would have its day.

But to return to Monterey. There are Titan talks all day—lab experiments to make tholin (the organic gunk of haze particles), new observations from large telescopes, such as the Keck and VLT (the Very Large Telescope in Chile), investigating Titan's surface, more ground-based work measuring Titan's winds, observations from space telescopes (not just HST but also FUSE, the Far Ultraviolet Spectroscopic Explorer), and theoretical studies of winds, the haze, and convection.

Even the lunch break is taken by Titan—a discussion to begin coordinating ground-based observations of Titan to supplement *Cassini*'s activities. In the evening, the conference hosts a buffet at the Monterey Aquarium, which has a marvelous jellyfish exhibit. It does not escape the notice of the nerdy astronomers that some of the delicate creatures look like spaceships. Next morning I head off to the airport and home. On the way, perhaps stimulated by the talks, perhaps because I am next to the sea in a rental car that reeks of hydrocarbons (see chapter 7), I think of Titan.

MILTON KEYNES, ENGLAND, OCTOBER 31, 2003

I awake, not too badly jet-lagged, to the quack of ducks. They are making their morning rounds in the canal a few feet from the window of my room at the Peartree Bridge Inn in Milton Keynes, England. Milton Keynes, somewhat infamous in England as a sterile "New Town," is home to the Open University (OU) and headquarters to the Huygens Surface Science Package team, PI John Zarnecki and his colleagues, having recently moved there from the University of Kent.

After breakfast I walk the mile or so to the OU, past a small stone church and some cows grazing quietly on this foggy morning. There is the sound of traffic not too far away. A modern rectangular grid of roads and roundabouts gives most people their unfavorable impression of Milton Keynes, but this network was simply draped over some English countryside with its farms, village pubs, and churches without obliterating all of it, retaining a palimpsest of the rural within the urbanization.

At the OU I meet with my old colleagues, some of whom I have worked with for over a decade. We discuss a number of tasks that need to be tackled in the fifteen months or so to Titan arrival. *Cassini*'s long cruise has seen many of us go bald, gray, or pudgy—but at least there is time to reconsider some of the settings of the experiments on the probe. We have a meeting to change one of these: the impact level that will tell the probe it has landed, whatever it turns out to land on. For some reason, a decade ago it had been set to some value, and no one remembers why. Looking at some old splashdown simulations I had made, I suggest we reduce the threshold. The change is approved and will be broadcast up to the probe at the next in-flight checkout some months ahead.

ARIZONA, NOVEMBER 6, 2003

A different sort of commute this week. It is 6:00 a.m. and I am driving down to Tucson from Mount Lemmon after a night at the Kuiper 1.54-m telescope. This telescope, named after the discoverer of Titan's atmosphere and the founder of the University of Arizona's Lunar and Planetary Lab, is better known to everyone who has used it as "the sixty-one-inch." Built originally to map the Moon for the *Apollo* missions, it is now rather small compared with frontline astronomical facilities but is still useful. I am working with an undergraduate student to monitor Titan, Uranus, and Neptune, watching for cloud activity with a spectrometer mounted on the telescope. This kind of work needs more nights than can realistically be obtained on the big telescopes. The sixty-one-inch is a nice compromise between the easy access needed and the size of the scope.

The drive between Tucson and the telescope takes about an hour. After climbing a few thousand feet, the saguaro cactus of the desert give way to scrub. The granite mountain is rather more bare than it once was after the devastating wildfires earlier this summer. Some grassy shoots give the blasted

terrain a greenish tinge, but the wiry black corpses of trees
are a stark reminder of how it once was. But presently we
climb above this area and are among undamaged aspen. Fire-
fighters managed to save the Catalina Observatory—the sixty-
one-inch, the adjacent Schmidt telescope used for hunting asteroids and comets, and the cozy dormitory.


New Mexico, November 7, 2003

I tread carefully as the smooth surface gets precipitously steep
near the rim. But the view from the rim is rewarding: across
the plain I can see a half dozen of the brilliant white dishes
of the Very Large Array (VLA). I am standing inside the
25-m dish of one of the antennas, parked in a vertical position
for servicing. Fourteen years before, the first radar echoes
ever received from Titan bounced from the parabolic metal
surface I am struggling to stand on into the radio receivers
at its center.

I have visited the National Radio Astronomical Observatory in Socorro, New Mexico, to give a colloquium on Titan
at the invitation of Bryan Butler. Bryan, who was a PhD student at Caltech when he worked on those first Titan radar
echoes in 1989 with Dewey Muhleman, is now an astronomer
here and managed to get me a visit into one of the VLA antennas. I have taken the visitor tour here at the VLA before, and
indeed often see the white antennas arranged in a Y pattern
from the air on the way to Europe from Tucson, but to get to
clamber on and in one of the antennas is a special perk. A
couple of years previously, Bryan had been a consultant to,
and indeed appears fleetingly in, the movie *Contact*. (Look
out for an equation being jotted down in the movie—that's
Bryan's hand, and the equation has nothing to do with
radio astronomy!)

We discuss some old radio astronomical observations Bryan
has made. Better publish them soon: only a year to go before
*Cassini* does much better than the array of telescopes around
us. *Cassini*'s crescendo of activity continues to build. For an
hour and a half before my colloquium, I had to borrow an

office to join one of our endless Titan observation planning teleconferences.

On the drive back to Albuquerque airport—was it only this time last week I was in Milton Keynes?!—I reflect on all the fascinating aspects of Titan and all the wonderful tools we have for exploring. What a way to make a living!

......................................................................................................................

## TITAN LOOKS LIKE MARS!

Dozens of diverse investigations went on in the years leading up to *Cassini*'s arrival. At the time, the telescopic monitoring, theoretical models, laboratory work, plus all the planning work on the project itself occupied only a small cadre of the planetary science community. But once *Cassini* arrived, Titan would become the "planet du jour," and many more people would get involved in analyzing *Cassini* data. Some of these pre-*Cassini* studies would form the foundation of all the subsequent work. Some would turn out to be utterly irrelevant.

Athena Coustenis hosted a Titan workshop in Paris in January 2004 to gather and summarize current impressions before *Cassini* arrived. Some smaller workshops had taken place in Berkeley and Boulder in the previous year.

What stole the show were the images that Antonin Bouchez and colleagues had acquired with the Keck. The map made at two microns was quite astounding. It looked like Mars! There was a long, dark H-shaped region near the equator, faintly reminiscent of the giant canyon system Valles Marineris on Mars, and a bright cluster of clouds around Titan's south pole, not quite resolved, looked like the Martian polar cap!

One key feature of this new map was that the contrasts at two microns were high enough not to need the clever background subtraction that had to be done a decade earlier to make the HST maps of Titan at shorter wavelengths where the haze was thicker. This meant that variations of brightness with latitude were real surface features. On the HST maps, where a "synthetic" average atmosphere had been subtracted, there was no way to discriminate a bright or dark band on the surface from a band of cloud.

Everyone noticed how all the darkest spots, like the H, were concentrated around the equator. Dark areas, it was thought, were probably where the liquids would be. Why might they be concentrated at low latitudes? It didn't make much sense, because there were suspicions that ethane might be preferentially deposited at high latitudes. It was also known that the clouds being seen in the present season were also at high latitude, in midsummer. So if it was raining anywhere, it was at high latitudes. Maybe Titan was egg-shaped, and liquids flowed on the surface, downhill to the equator. It was weird. The truth would turn out to be stranger still, and quite unexpected.

# 3. *Cassini* Arrives

It's July 1, 2004, and *Cassini*'s long trek to its destination is over. In just one hour and fifty-two minutes, the spacecraft will be closer to Saturn than at any other time during its mission. The engine burn that will curtail *Cassini*'s interplanetary trajectory and deflect it into orbit around Saturn—Saturn Orbit Insertion, or SOI—is only minutes away. This burn will be the most critical event of the mission since launch. But first the spacecraft is crossing through the ring plane. To minimize the risk from a collision with a ring particle, it makes its passage through the rings in the large gap between the F and G rings. Shortly after *Cassini* has crossed above the rings, the main engine burn begins at 01:12 Coordinated Universal Time (UTC) and continues for ninety-six minutes.

........................................................................................................

### RALPH'S LOG, JULY 1, 2004

I spent part of the morning filming with the BBC, driving up and down Pasadena's Colorado Boulevard pretending to "commute," and then pretending to check my e-mail on a laptop in a motel room (not my own). Just like the hundreds of contingency plans, backup sequences, or observation designs that don't get used on a space mission, the media take a lot more footage than they actually use, and thankfully this pedestrian stuff fell onto the cutting room floor.

Saturn Orbit Insertion would be somewhat tense. A couple of Mars missions had been lost at just such arrivals, when fuel pipes exploded or bad navigation hurled them into the planet. In the emptiness of space, spacecraft are usually quite safe, but maneuvering in close proximity to a planetary body is a risky time. This is the moment when controllers find out whether the rocket motor will fire properly. For *Cassini*, there was also the apparent hazard of crossing the ring plane.

The media, with the space agency public relations machine as a willing accomplice, talked up the risk, and cameras zoomed in on the tense faces of controllers. In reality, there was little to control—seventy light-minutes away, whatever was going to happen was going to happen, and no heroic engineer was going to grab a joystick and be able to swerve *Cassini* out of trouble, even if we knew that some rogue ring particle loomed in our path.

And so, everything went just the way it was supposed to—pretty much exactly as portrayed in the drawings in the Phase A report back in 1988. *Cassini* passed across the ring plane, fired its motor for the required time, and ducked back across the rings. The only information we got was a radio tone, shown as a sequence of points on a huge screen. A green diagonal line showed what was supposed to happen, which was the frequency of the radio signal suddenly starting to change as the changing speed of the spacecraft during the engine burn altered the amount of Doppler shift. Every so often, as expected, the received data points would diverge from the "predicted" line as the denser parts of Saturn's rings blocked the signal. But then the signal would come back, to cheering. And finally, when the green diagonal line straightened back to horizontal at the end of the motor burn, the red data points followed and did just that. Although some of the cheering and high-fiving seemed a bit forced, the relief and excitement were nonetheless genuine.

Saturn had a new satellite. We had arrived.

..............................................................................................

*Cassini* was on the brink of an unprecedented tour of exploration and discovery. But the saga that had successfully brought *Cassini* to this point,

Figure 3.01. Artist's impression of the release of the *Huygens* probe by *Cassini*. In reality, Titan would be much farther away, a mere dot. (NASA)

with its highs and lows, crises and triumphs, had already been going on for more than twenty years.

## THE IDEA FOR *CASSINI*

The *Cassini* story began in the early 1980s, soon after tantalizing *Voyager* encounters with the Saturnian system in November 1980 and August 1981. The European Space Agency called for ideas in 1982, and an international group of three scientists responded. Wing Ip of Germany, Daniel Gautier of France, and Toby Owen from the United States proposed a mission they called *Cassini*, which would be a Saturn orbiter carrying a Titan probe.

This concept was examined by NASA and ESA jointly. NASA's *Galileo* mission to Jupiter was carrying a probe, which would separate from the main spacecraft and descend through Jupiter's atmosphere in 1995. There was a spare of this probe, and the original notion was to use it as the probe to go down onto Titan; ESA would provide the orbiter. However, the roles were later reversed. NASA engineers designed a new generic spacecraft, or "bus," called *Mariner Mark II*. The idea was to save development costs by using the same basic vehicle for a series of missions.

The logic there was straightforward, but taking on the responsibility for a planetary probe was a big step for ESA, which had not built one before. The closest thing had been a comet flyby spacecraft, *Giotto*, in 1986. It would require important technological capabilities new to ESA, for entry shields, parachutes, and so on. An assessment study was undertaken in 1984. It was followed by a "Phase A" study, begun in 1987 after a delay of a year to synchronize the schedules and budgets of NASA and ESA.

By late 1988, the technical challenges involved in constructing not only the probe vehicle but also the instruments to be carried were sufficiently understood that it was possible to consider the probe as a real mission rather than just a study. In November 1988, subject to NASA approving *Cassini*, ESA selected the Titan probe to be funded and named it *Huygens*. Time was now of the essence. The preferred route to Saturn and Titan would use the gravity of Jupiter to sling the spacecraft on its way, and Jupiter would be favorably positioned for this gravity assist maneuver only between 1994 and 1997. The U.S. Congress approved the start of *Cassini* in 1989. The joint mission was on the road.

Teams of scientists on both sides of the Atlantic considered what experiments they could carry out with *Cassini* and *Huygens*. The experiments would be competitively selected— the stakes were high. At this time, planetary missions were elaborate and carried many instruments, but were few and far between. The "faster better cheaper" era introduced by NASA Administrator Dan Goldin lay in the future.

## INSTRUMENTS FOR *HUYGENS*

The *Cassini* project got going in earnest in autumn 1990, with the announcement of the selected payloads. Six experiments were chosen for *Huygens*:

1. The gas chromatograph/mass spectrometer (GCMS) led by Hasso Niemann of NASA's Goddard Space Center
2. The descent imager/spectral radiometer (DISR) led by Marty Tomasko of the University of Arizona
3. The Doppler wind experiment (DWE) led by Mike Bird of the Radio Astronomy Institute at the University of Bonn

4. The aerosol collector/pyrolyzer (ACP) led by Guy Israel of the Service d'Aeronomie, Paris
5. The Huygens atmospheric structure instrument (HASI) led by Marcello Fulchignoni of La Sapienza University in Rome, Italy—later at the Paris Observatory in France
6. The surface science package (SSP) led by John Zarnecki, formerly at the University of Kent, but later at the Open University in the United Kingdom

At 18 kg, the GCMS was the most massive instrument, and in many ways the most important experiment. It would analyze the chemical composition of Titan's atmospheric gases and of gases liberated by the ACP (discussed later). The GCMS would be able to distinguish between and identify the great variety of compounds in Titan's atmosphere, not only by their molecular mass but also by their affinity to special coatings inside thin tubes, along which different compounds take different times to travel. It was a complex instrument to tackle a difficult job.

The DISR was *Huygens*'s camera that would take pictures looking both directly down and to the side as the probe descended. But in addition, it had the ability to record the spectra of sunlight filtering through the haze and reflected up from the ground. It could also measure the sunlight scattered around the Sun (the aureole), to obtain data about properties of the haze.

The DWE was one of the simplest. The only hardware for it on the probe was an ultra-stable oscillator on one of the two channels of the probe's radio link. The principle behind it was to measure the change in frequency of the signal received by the orbiter by comparison with a reference oscillator on the orbiter. From this it would be possible to determine the probe's motion. Of the measured Doppler shift in frequency, most would be the result of the orbiter's rapid approach toward Titan, and some would be due to the probe's vertical descent. These two components could be calculated and removed. Any remaining component of the probe's motion would be due to wind.

The ACP was a novel instrument designed to trap aerosol particles by sucking atmospheric gas through a tiny filter held out in front of the probe. The filter would then be pulled inside the instrument and heated in a set of temperature-controlled ovens. Aerosol material breaks down when heated. The gaseous products released were to be transferred through a pipe to the GCMS instrument for analysis.

The HASI was to measure the basic properties of Titan's atmosphere, such as pressure and temperature, during the probe's entry and descent. A capability for HASI to digest the signal from the probe's engineering altimeter was added later.

The SSP was a collection of small sensors. As well as an accelerometer and penetrometer to characterize an impact on a solid or soft surface, the package included instruments to take measurements in the event of a "splashdown" in liquid. Tilt sensors would measure the probe's orientation and bobbing motion in any waves. An ingenious refractometer would measure the ethane/methane composition of the liquid, with additional information provided by thermal, density, and electrical permitivity sensors. An acoustic sensor would act as a sonar to measure the depth.

The hope was that together, the six experiments would maximize the data return during the few precious minutes of *Huygens*'s descent, within the constraint of the size and mass capacity of the probe.

........................................................................................

### RALPH'S LOG, AUGUST 1994

Each of the instruments involved the efforts of dozens of scientists, engineers, technicians, and students. One of them was me.

One function of the SSP was to measure the mechanical properties of Titan's surface by recording the impact of the probe, if it survived. Since I had a background in aerospace engineering, as well as a detailed knowledge of the probe, having worked in the ESA project team for a year already, this investigation seemed like an ideal fit with my interests and expertise and would make a good PhD project.

It is rare that such a junior individual gets to play such an identifiable role in a major space endeavor, but someone has to do it. The budget for the SSP was tight, and as a PhD student, I was cheap labor. So, at the tender age of twenty-two, I was made responsible for specifying what sensors SSP should use—how fast and how accurately they would have to be recorded, and so on. There had been preliminary designs, of course, but these had been proposed without knowledge of the final probe design.

Between 1991 and 1994, the SSP team decided that two sensors would make the impact measurement. For very soft landings, or splashdown into a liquid, an accelerometer mounted on the SSP electronics box would do. I had read up on all the details of splashdown testing from NASA records of the *Apollo* program. That bit was easy. An off-the-shelf accelerometer would be used, rather like those used to actuate air bags in cars. But for impacts into harder materials, the acceleration recorded would be affected more by the structural properties of the probe itself than by the surface hardness we wanted to measure. And so I designed a penetrometer, a force sensor that would stick out of the bottom of the probe on a rigid mast. If the probe landed flat, this would be driven into the ground, generating a force signal.

We had to check that the sensor would work after a lethal dose of radiation, such as the one the probe would get flying by Jupiter, and that it would function acceptably at Titan's surface temperature. A way had to be found of checking the instrument's health en route to Titan—these and dozens of other details were worked out on paper before it could actually be built.

Eventually, the design details were settled and all the parts arrived—titanium alloy for the penetrometer head, the custom-made piezoelectric ceramic disks. Even the wires in SSP were special: stainless steel instead of copper, to minimize how much of the probe's warmth would leak out along them.

And so, here I was one summer day in the University of Kent's Electronics Workshop, feeding a little wire through a little hole. I held it in place as Trevor Rees applied the soldering iron and bonded the wires to the transducer, ready for their long, cold trip. (I was not permitted to do the soldering myself—that had to be done by someone who had been space-certified after taking a special ESA course in high-reliability soldering.)

In fact, we built five penetrometer heads. The best one, the one where the wire seemed to bend a little more smoothly, where the solder joint seemed just that bit neater, was designated the flight model—the one that would go to Titan. Another two, almost as good, were also double-bagged and placed in the secure "flight cupboard" as flight spares, in case something were to go wrong. The other two, which worked fine but weren't quite as sound, would be used in lab tests and student projects.

....................................................................................................

## HUYGENS: DESIGNED FOR ITS JOB

The design of the probe was to deliver a payload of 50 kg of instruments into Titan's atmosphere and have them descend to the surface in about 135 minutes, while keeping them warm, supplied with power, and transmitting their data to *Cassini*. Additional requirements were that the probe should start its descent around 160 km altitude, that it notify the experiments when the altitude dropped below 10 km, and that the probe spin at a few revolutions per minute during descent so the optical instruments could pan around. Further, it should support the payload for a minimum of 3 minutes after impact. However, survival beyond impact was not a requirement. There was no way it could be, without knowing more about Titan's surface.

In order to do all that, of course, it would also have to survive the noise, acceleration, and vibration of being launched into space on a rocket, spending seven years in the vacuum and radiation of space while attached to *Cassini*, and another twenty-two days coasting by itself. But perhaps the biggest challenge of all, the toughest one to be sure about, was entry. The twenty-two-day coast sounds benign, but what this means is hurtling at 6 km per second toward Titan. At this speed, the kinetic energy of the probe is the same per kilogram as a high explosive like TNT. Somehow the probe would have to shed that energy.

To reduce its speed, the probe was made with a round-nosed conical airbrake or decelerator. This would act like a rigid parachute, slowing the probe down by air drag. But at hypersonic speeds, the air would glow with the violent energy of entry, and a bare structure or parachute would

Figure 3.02.  A schematic representation of the *Huygens* descent sequence. After hypersonic entry in its heat shield, a pilot chute pulled off the back cover and allowed the main chute to inflate. This slowed the probe down and allowed the front shield to fall away; thereafter, the main chute was released and the probe descended under a small "stabilizer" parachute. It is pictured here finally operating on the surface. (ESA)

melt. So the decelerator was coated with a high-temperature insulator or heat shield.

The prime contractor for the probe, selected through competitive bidding, was Aerospatiale of Cannes, France (later to become part of Alcatel and finally, as Europe's post—cold war aerospace industry consolidated, Alenia-Alcatel Space). It headed a consortium of suppliers from several European countries, including Germany, Spain, Italy, and the United Kingdom. As with other European Space Agency projects, the industrial contracts had to be shared among the member states, the value of the contracts going in rough proportion to the contributions each country makes to ESA's budget. On this basis, France, Germany, and Italy were to take the lion's share of the work, followed by Britain, Spain, and the many smaller countries. It is a challenge for the project managers in ESA, who are usually based at its technical center, ESTEC, in Noordwijk, the Netherlands, to juggle the bids from various companies to make sure that France is supplying 35 percent of the project, and Denmark its 2 percent, and so on. It doesn't necessarily lead to the most efficient technical solutions, but it is a system that works surprisingly well, bearing in mind the diversity of ESA's members.

Aerospatiale would lead the technical design and be responsible for accommodating the experiments. They would also supply the all-

Figure 3.03.    An engineer in a white clean-room suit performs some final assembly tasks on the *Huygens* probe at Cape Canaveral. The front shield is visible behind the white support fixture. Note the cold air hose in the foreground. (NASA/KSC)

important heat shield, based on tiles of a silica material used in French ballistic missiles. Alenia and Laben of Italy provided the data handling and communications systems. A British company, Martin-Baker, developed the descent control system, the set of pyrotechnics, lines, and parachutes that would pull the probe away from its heat shield and bring it safely down to the surface in the allotted time. The metal structure of the probe was to be supplied by CASA of Spain.

A spacecraft, like any other modern aerospace project, is not just hardware. Important elements included the onboard software and the procurement services to secure the space-rated pedigree of its components (provided by Logica and ITT in the United Kingdom). The probe itself would be assembled in Germany, where extensive tests would take place to verify its ability to function in the vacuum of space and in the deep cold of Titan's atmosphere, to check its ability to tolerate lightning strikes, and so on.

All of this, as well as the important interfaces with the *Cassini* project in the United States, had to be coordinated. Literally hundreds of face-to-face meetings and teleconferences would negotiate and agree to the various technical and contractual details—all, of course, to be recorded

Figure 3.04. The day-to-day business of spacecraft and instrument engineering is not as surgical as figure 3.03 implies. At left is the *Huygens* probe "engineering model," the probe equipment deck festooned with wires to verify electrical function. The metal structure is not present. At right is a test of the DISR instrument's Sun sensor on the roof at the University of Arizona. (ESA and R. Lorenz)

in tons of thick documents, changes to which would be meticulously tracked. And almost any detail or function on a spacecraft has to be tested. Sometimes designing and executing the test is as much of a challenge as designing and building the thing in the first place! It was a great deal of work. At the beginning, in 1990, much of the communication was by fax; e-mail was not as prevalent as it is today, nor had Microsoft Office software yet emerged as a global standard. And the World Wide Web did not even exist! But this large-scale system engineering activity and project management process—arguably a more significant spin-off of space exploration than the nonstick frying pans that are often (erroneously) cited—brought it all together in the end. Time zones, computer incompatibilities, different units of measurement, different sizes of paper, different languages and ways of doing business are all just barriers to be overcome by patience and hard work.

Beyond the cultural differences within Europe and between Europe and the United States, another cultural divide must be overcome on a project like this—namely, that between scientists and engineers. Although both are technical disciplines, they have very different styles. Engineers are more team-oriented, used to the hierarchical organization of engineering organizations. Scientists are more often prone to prima donna behavior but sometimes less constrained in how they work.

Many people trained in science go on to engineering careers; rather fewer people go the other way. The scientist as the "customer" wants to supply a massive, power-hungry instrument that looks in all directions and generates a mountain of data that the scientist and his or her colleagues and successors can sift through for years afterward. The engineer, with similar demands from all directions and faced with ultimate limits on the cost of the project, the size of the launcher, and so on, must work to keep the payload and its needs manageable. The engineer's job would be easiest if a scientific instrument were just a small box that could be put anywhere on the probe, and didn't need any power or data. The dynamic tension between the two disciplines results in intense negotiations but contributes to an optimum solution emerging.

A peculiarity of ESA's system (compared with NASA's) is that the agency does not pay for the instruments that fly on its missions. The costs of both hardware and scientific personnel are paid for by individual member states and consequentially can be subject to all kinds of uncertainty. In contrast, NASA pays for both the platform and the instruments, and thus has some measure of overriding control.

Although the *Cassini* orbiter was for the most part built in the United States, and the *Huygens* probe in Europe, and their payloads likewise, there were components of each from the other side of the Atlantic. Some components on the *Huygens* probe, like batteries and accelerometers, came from the United States, but the orbiter's high-gain antenna was built in Italy. On the scientific side, the two most complex experiments on the probe, the gas chromatograph/mass spectrometer and the descent imager/spectral radiometer, were led in the United States, while the teams working on the *Cassini* dust analyzer and the magnetometer on the orbiter were led in Europe. *Cassini—Huygens* was therefore a monumental joint effort, requiring patient coordination.


## *CASSINI'S* TOOLS FOR TITAN


In parallel with the probe's development taking place primarily in Europe, the much larger orbiter spacecraft was being put together in the United States. The instruments selected for *Cassini* were announced soon after the list for *Huygens*. They included several that would be used to study Titan during the dozens of close flybys it would make through the course of the mission.

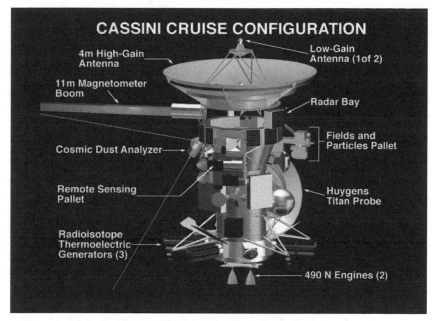

Figure 3.05. Configuration of the *Cassini* spacecraft. (NASA/JPL)

Sadly, the contracting economy of the early 1990s took its toll on plans for planetary exploration. The "two for the price of one and one-half" logic of *Mariner Mark II* faltered against a NASA budget crunch in 1992 as NASA struggled to pay for its part of the International Space Station. And so *Cassini*'s sister mission, CRAF (Comet Rendezvous and Asteroid Flyby) was canceled. *Cassini* survived this financial crisis by the skin of its teeth, chiefly because of its international nature, which would have made cancellation look very bad. But *Cassini* was slimmed down, many of its instruments having their capabilities reduced ("descoped"). Crucially, the scan platforms—devices to point instruments independently of the body of the spacecraft—were deleted to save money. By any rational life-cycle costing, this would be seen as a false economy, the savings in construction in 1992 being eaten up and more by the additional operations complexity years later because the whole spacecraft would have to be slewed around to point its different instruments at targets as it whipped by.

Even with descoped instruments, *Cassini* had a formidable science payload. Disentangling the bewildering array of phenomena in the Saturnian system—the interactions of the rings, the satellites, the magnetosphere,

and the atmospheres of Saturn and Titan—requires simultaneous measurements of many kinds. It is not the sort of exploration that can be
tractably addressed with a small mission.

*Cassini* was the first outer solar system mission to be equipped with
radar. Unlike most radar-mapping spacecraft, which operate over a fairly
limited range of altitudes, *Cassini* would be making radar observations
from as far away as tens of thousands of kilometers to within 4,000 km
of the surface. To cope with this, the design needed to be highly versatile.
All of Titan was to be mapped with low resolution, but in addition, strips
totaling about one-quarter of the entire surface would be charted much
more finely. In sweeps during *Cassini*'s close passes over Titan, the radar
would look to the side with five beams that form a line. Dragging this
line across the surface covers a strip a couple of hundred kilometers across
and several thousand kilometers long. The signals are processed in a
clever way so that pixels only 400 m across can be measured.

The visual and infrared mapping spectrometer (VIMS) was to be another versatile instrument with applications to Titan. Its field of view is
a narrow strip, but each strip is turned into a high-resolution spectrum.
By sweeping the strip around, or by simply letting the spacecraft's motion
drag the strip, an image cube can be built up—a stack of images of the
same scene in many hundreds of different colors. Effectively, VIMS records the spectrum of many pixels in an image simultaneously.

*Cassini*'s imaging instrument, known as the ISS (imaging science sub-
system), consists of two telescopic cameras—one wide-angle and one narrow-angle. Both have sensitive CCD detectors and an array of filters.
Some of the filters have been specially matched to the 0.94-micron "window" in Titan's atmosphere, to probe Titan's surface.

A key instrument for Titan was the ion and neutral mass spectrometer
(INMS). Even though *Cassini* would be flying 1,000 km above Titan's
surface, there would still be enough atmosphere at that altitude for its
composition to be measured by this instrument, pointed forward like a
scoop, literally counting the molecules of different masses.

The ultraviolet imaging spectrometer (UVIS) would be used to detect
airglow—the means by which *Voyager* had detected molecular nitrogen
at Titan. It would also be used in occultations of the Sun and of stars to
profile the amount of gas and haze in the upper atmosphere.

As the *Galileo* spacecraft did on its mission to Jupiter, *Cassini* carries
a set of sensitive magnetometers. Water, especially if it is salty, conducts

electricity—unlike ice and rock, which do not. Moving through a giant planet's magnetic field, a subsurface ocean on a moon generates a small magnetic field of its own. *Galileo* detected the appropriate magnetic signature for Europa as expected, but when Callisto showed the same magnetic signature, it came as something of a surprise. *Cassini*'s magnetometer would be able to carry out a similar test on Titan.

*Cassini* was equipped with a radio communications and tracking system more elaborate than any other planetary spacecraft has ever had, and it would be possible to probe Titan's atmosphere by means of its signals. Passing the radio signal through the upper atmosphere of Titan could determine the density of free electrons. Lower down in the atmosphere, the signal is refracted by the denser layers. The radio signal is also affected by Earth's ionosphere and the tenuous gas between the planets. However, by measuring its three frequencies simultaneously, these effects can be removed to get much more accurate measurements. *Cassini* also carries long, sensitive radio antennas to search for electrical phenomena in the Saturnian system and to listen for lightning at Titan.

## LAUNCH

*Cassini* was launched from Cape Canaveral in the early hours of the morning (local time) of October 15, 1997, by a *Titan IV/B* launch vehicle, the most powerful launcher in Western inventory. Just as sailors must take the tide or remain stranded, planetary exploration is governed by windows of opportunity, when planets are aligned correctly. *Cassini* had rather stringent constraints on its launch, since it relied on no less than four planetary encounters to hurl it on to Saturn. The launch window would last only a month or so. If there were a problem that took longer than that to fix, the launch would have to slip by months but the arrival at Saturn would be delayed by years, and fuel that could be used at Saturn would be consumed in getting there.

The probe and its experiments, shipped from Germany, had already been in Florida for months, undergoing various interfacing and compatibility tests with the orbiter, trucked over from California.

A last-minute hiccup did delay the launch. *Huygens* was well insulated against the cold of space and could get hot during tests on the relatively warm Earth. So a duct blew cold air continuously into the probe to keep

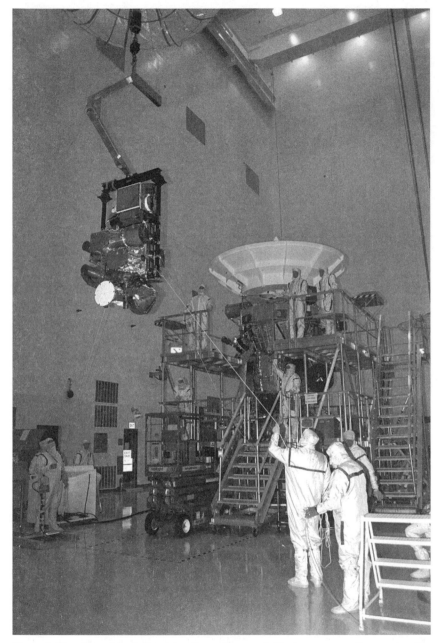

Figure 3.06.   Assembly of *Cassini* at Cape Canaveral. The instrument pallet carrying the cameras and spectrometers, wrapped in gold-colored Kapton insulation, is being lowered toward the *Cassini* orbiter structure in the scaffold. (NASA/KSC)

it cool. However, somehow the airflow on this duct got set far too high, and the gale-force blast of air shredded some of the insulating foam on the inside of the probe. The risk that particles of foam might have penetrated some sensitive component couldn't be taken. The probe had to be removed from the launch pad. Opening the probe up and vacuuming its insides took nearly a couple of weeks, half the available window.

After this slip, launch was rescheduled for October 13. A large crowd of scientists and engineers gathered in the early hours to watch the launch. But, as can often happen, a computer glitch and high winds at altitude prevented the launch from taking place. There would be another two-day delay.

In fact, although one wouldn't want to do this as a matter of policy, in case the weather turned sour or some mishap occurred on the launch pad, the two-week delay in launch actually had a major benefit. The launch window represents a period during which conditions are acceptable—the planets are in the appropriate alignment and so on. But not all moments in the window are the same, the middle of the window usually being closest to optimum. And so it was with *Cassini*. The mission would require more fuel to reach Saturn if launched at the beginning or end of the window. The delay caused the launch to take place in the middle of the window, when the fuel costs were less. This was good news. If the rest of the mission went without major problems, the fuel saved could be used to prolong the mission, or to perform a more aggressive tour with more close flybys of Titan.

Fortunately, on the fifteenth of October 1997, *Cassini* got off the ground safely. This night launch was quite spectacular. Not only was the Titan rocket the largest conventional launch vehicle in U.S. inventory at the time, but the monster solid rocket motors strapped to its side have an extremely luminous yellow exhaust. Even from miles off, the ascent was brilliant, the rocket itself invisible next to the dazzling columns of flame. Nature added an aesthetic flourish, a fluffy cloud over the launch pad— not a thunderstorm, thankfully, which would have scrubbed the launch. As *Cassini* ascended through it, the cloud was lit up from inside like a Chinese lantern. It seemed a good omen.

The roar of the engines took many seconds to reach the assembled spectators, presaged by a bit of a rumble that propagated through the ocean. Thus, the engine ignition took place in apparent silence, soon replaced with cheers and applause. It was a cathartic moment, a huge relief.

Even the most trustworthy rockets are only 97 percent reliable, and the upgraded solid motors on Titan, which would save *Cassini*'s fuel for use at Saturn, were comparatively untried. But everything worked perfectly, and after a minute or two, the burned-out husks of the solid motors tumbled like comets toward the sea, their job done, while *Cassini* sailed off into the darkness.

The scientists gathered at the launch site still had several days of meetings—to thrash out the design of the tour, the specification of archive data products, the latest updates to planning software, and so on. But the launch was a watershed in the project, marking the transition from hardware to operations. Many engineers would move on to new jobs. And although some of the scientists would remain busy in planning the tour, calibrating the instruments in flight, designing observations, and so on, many would disappear from the scene for a few years until their expertise was called on to analyze *Cassini*'s data.

························································································

## RALPH'S LOG, 1998

### DS-2

Although I was heavily involved in the various planning tasks, *Cassini* was not enough to support my research full-time, and so I pursued some other opportunities. A project that made a neat fit with my *Huygens* work came up in early 1998. This was a small Mars Penetrator project called the DS-2 Mars Microprobes, part of NASA's New Millennium Program of technology validation missions, the idea being to alleviate the risk in trying new technology on big science missions by showing they work in fast, cheap projects first. At $28 million, it was a tiny fraction of the cost of *Cassini*.

The two microprobes were like tiny versions of *Huygens*—about the size of a basketball. They would piggyback to Mars on the *Mars Polar Lander* (MPL) mission, to arrive in December 1999. Each microprobe weighed about 4 kg and was encased in a lightweight but brittle heat shield. There was no parachute; these probes would slam into the ground at four hundred miles an hour. But they were designed for that—the

Figure 3.07.   Launch! *Cassini* blasts off on the *Titan IV* launcher in October 1997, a spectacular night launch. (NASA)

heat shield would shatter, and these specially toughened probes would penetrate about half a meter into the ground, decelerating at some 20,000 g. There they would cook a tiny sample of Martian soil to see how much water was locked up in the ground (this was, of course, long before the *Mars Odyssey* spacecraft discovered lots of permafrost remotely in 2002).

My role in the project was to determine how we could measure the ice content of the ground, find any layering, and measure the depth of penetration using the deceleration profile recorded at impact. This was jolly good fun, and involved blasting mock-ups of the probes into the ground with a huge air cannon on the back of a truck at a ballistics facility in New Mexico, with dead helicopters and tanks with holes in them littered around. Many of the technical problems that had to be confronted were similar to those on *Huygens*, but the style of the small project was very different, and the schedule almost ridiculously fast.

Martian scientists have it easy. There is almost instant gratification: you build your instrument or spacecraft, launch it, and in less than a year, you are there, getting data. After a smooth launch in January 1999 (a daytime launch on a small Delta rocket—impressive, but not quite as awesome as *Cassini*'s nighttime Titan), MPL and its two diminutive sidekicks, since named Scott and Amundsen in a student contest, neared Mars on December 3, 1999. The DS-2 probes would last only a couple of days before their batteries ran out, and the DS-2 science team geared up to present some first results at the AGU conference the following week.

JPL (the Jet Propulsion Laboratory) was a zoo. Even the fact that much of the MPL science operations would be performed from UCLA a few miles away did not seem to take the heat off—landing on Mars is big news! Scientists had to use the remote wilds of the car park because the prime spots were taken by VIPs and television trucks. Suddenly the policy of not tailgating someone through the access-controlled door to the operations building was enforced by a humorless guard with a gun. We got ready, briefed on the best color scheme to use in plots of data so that they show well on TV. We gave interviews, discussing what we hoped to find. The data would arrive in the early evening, so we tried to get some extra sleep ahead of time so we could work through the night.

And then, at the appointed hour, silence. Nothing. No signal from either of the microprobes, or worse, from MPL either. Well, maybe the battery was cold, and the spacecraft will report in during the next window in a few hours. No such luck. Engineers run through the various scenarios—well, if this failed, then this would switch in automatically eight hours later, and the lander should respond to ground commands. The scenarios played through, but still no response.

After a couple of days, we just had to give up and go home. We knew DS-2 was a risky mission and surely expected that one might fail. But to lose both, and MPL as well, was a horrifying shock. No exciting new findings, no demonstrating new, efficient ways of delivering payloads to the Martian surface. Nothing but some test reports and some lessons learned. That's show business.

God, I hope *Huygens* doesn't end up like this.

........................................................................................................................

## PLANNING THE TOUR

It would take seven years for *Cassini* to reach Saturn, but even that interval would be too short to work out all the details of what to do once it got there. First the teams had to agree on the "tour," the best orbital path around the Saturnian system. The nominal length of the mission was four years, and during that time, *Cassini* would make over seventy orbits of Saturn. The challenge was to devise a tour that would optimize the opportunities for all the experiments on board to gather observations of their targets, whether Saturn itself, the rings, or moons. And of course, early on in the tour, there would be the release of *Huygens*.

Once that was agreed (and trying out dozens of possibilities, progressively reaching a better design, took over five years), then the minute-by-minute sequence of pointing the spacecraft and designing the observations themselves (exposure times for the cameras, bandwidths for radio instruments, and so on) had to be done. This process was, of course, made all the harder by the compromise in *Cassini*'s design that had to be made in 1992, the deletion of the scan platforms. Instead the cameras were bolted to the side of the spacecraft, which meant that pointing them would require the whole spacecraft to be turned around. It also meant that it would not be possible to point the antenna at Earth while images were being taken. Operations would have to be arranged to take place sequentially rather than simultaneously. So even as the hardware moved to Florida in the latter half of the 1990s, and then into space, there was plenty of work to do.

Though *Cassini* left the launch pad in October 1997, it would not actually leave the inner solar system until almost two years later. Even the most powerful rocket available, the *Titan IV/B*, was not capable of imparting enough speed to *Cassini* to propel it to Saturn. *Cassini* was a monster of a spacecraft, at 6.8 m long among the largest ever launched and weighing in at 5.5 tons. The solution was the gravity assist technique—picking up speed in a close encounter with a planet to create a kind of slingshot effect. *Cassini*'s flight plan required not just the close flyby of Jupiter that constrained the overall window of opportunity for the mission, but three helping hands in the inner solar system. So it was that *Cassini* headed first toward Venus. In two complete orbits around the Sun in the inner solar system, *Cassini* swung past Venus twice, on April 26, 1998, and June 24, 1999, and then once by Earth, on August 18, 1999. Passing Earth at a distance of 1,180 km added a valuable 5.5 km/s to *Cassini*'s speed, sending it on its way at about 25 km/s.

## PUTTING THE PETAL TO THE METAL

Once at Saturn, there would be a near-infinity of options. In the broadest terms, the first order of business was to deliver the probe and get the orbital period down by bleeding off energy at the first few Titan flybys—rather like the slingshot technique in reverse. Then successive Titan encounters would use Titan's gravity to slowly change *Cassini*'s orbit around Saturn. At each flyby, the orbit could be changed in one of several ways. The orbit could be shrunk or grown, changing its period—usually from one resonance with Titan's sixteen-day period to another, such as from a thirty-two-day orbit to a sixteen-day one, since *Cassini* would have to reencounter Titan to make further changes. Or the orbital plane could be changed—raising or lowering the inclination. Or the orbit could be rotated within its plane, the so-called petal rotation.

The architecture that evolved, after literally dozens of trials, was as follows. After the first couple of orbits, during which the *Huygens* probe

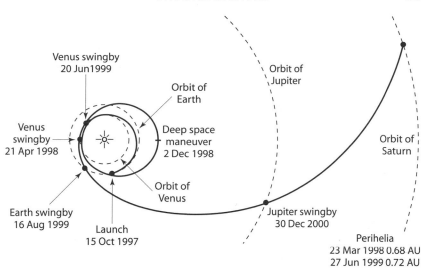

Figure 3.08. *Cassini*'s interplanetary trajectory took it two and one-half times around the Sun, on four planetary flybys, before it reached Saturn. (NASA/JPL)

would be delivered so as to arrive with the Sun in the right part of the sky for the camera to work, and so on (the second orbit being backup for the first), the inclination would be brought down to zero, putting *Cassini* in the Saturnian ring plane. This would be good for several things, notably for many close encounters with Saturn's other satellites and an edge-on view of the rings, as well as occultations of Saturn's atmosphere. Then the petal would be rotated from the dayside to the nightside, permitting an exploration of the Saturnian magnetotail. After that, in what would be late 2006, the inclination would first be increased with orbits that encountered Titan as *Cassini* flew away from Saturn (outbound encounters), then decreased with inbound encounters. This novel maneuver, called a 180-transfer or "cranking over the top," flipped the orbit back into the dayside like a pancake. Finally, the orbital inclination would be progressively increased to allow *Cassini* to look down on Saturn's pole and the rings toward the end of the nominal mission, finishing up with an inclination of some 75° in June 2008.

All in all, there would be forty-four close Titan encounters, numbered T1 to T44. A less "Titanocentric" system also referred to the orbit or "rev" numbers; yet a third system referred to sequences, the several-week-long blocks of commands sent to the spacecraft. Thus, Titan flyby T8 occurred on rev 17, in sequence S15 on day 301—better known as October 26, 2005.

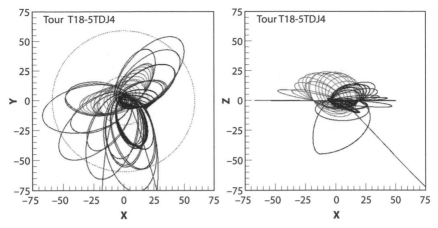

Figure 3.09. *Cassini*'s tour around the Saturnian system, designated T18-5. This particular variant is T18-5-JD4. The left view is from the north pole, with the direction to the Sun toward the bottom. The right view is from the Sun, north up, showing the inclination of the orbits varying throughout the tour. The lowermost, slightly kinked petal is the first orbit, the kink being the periapsis raise maneuver. (NASA/JPL)

In the original plan, *Huygens* was to be released on the first Titan flyby (T1), planned for November 27, 2004. However, the mission was redesigned (see the following section) to include an additional orbit around Saturn and for the probe to be released a little later, so it would reach Titan in January 2006, on a third flyby. Because so much work had gone into planning observations in the tour, and countless documents already used the existing sequence numbers (it is easier to reprogram a spacecraft a billion miles away than it is to reprogram a bunch of scientists on Earth!), the new tour began with flybys TA, TB, and TC before picking up at the old T3.

## A BIT OF TROUBLE WITH *HUYGENS*

During the Earth flyby, engineers took the opportunity to test out various instruments and systems. One of those tests was of the radio link system that would operate between *Huygens* and *Cassini* during the probe's descent. A signal was transmitted from Earth, simulating the probe. It soon became apparent that all was not well.

It emerged that a design compromise years before in the Europe-supplied radio receiver on *Cassini* that would receive *Huygens*'s signal led to it being highly sensitive to the exact rate at which it received the radio bitstream of data. If the received frequency were slightly off, the bitstream would become desynchronized and the data would be corrupted. Unfortunately, although the overall receiver design accommodated all the anticipated factors on the frequency and had been tested on the ground, the Doppler shift due to *Cassini*'s motion relative to *Huygens* was enough to trip up the bit synchronizer, and this rather difficult test had not been attempted.

When the results came back from the Earth flyby, the catastrophic problem was soon diagnosed. There were many places where the design could have been fixed, in hardware or software, but only when the probe was on the ground. The crucial parts in the radio were not remotely reprogrammable. A recovery team was convened, with experts in the mission and in digital radio design to hand. Although the situation looked bleak, the team members at least had time on their side. There were still five years to go.

The solution involved several changes to the mission. First, the probe would be switched on several hours early. It looked as if there was enough battery power to do so, since the batteries seemed to be holding their charge well. This would cause the radio transmitter to warm up, shifting its frequency slightly and improving the synchronization. The software on the probe, and on the various experiments, had to be modified accordingly. Some changes in ground software would recover a few of the otherwise-rejected data packets. But the biggest change was to the orbiter mission. In the original plan, *Huygens* was to be delivered on the first orbit. By delaying the delivery of the probe for a few months, and flying by at a greater distance, the Doppler shift due to the relative motion of the probe (dangling under its parachute in Titan's atmosphere) and the orbiter, whipping past at 6 km/s, would be reduced to a level where the bit synchronizer could lock on correctly.

Simulations of the radio performance showed under what conditions the receiver would lose data. One particularly instructive plot of signal strength against frequency showed a jagged region, the "shark's teeth," where it would fail, and tests on the spacecraft as it cruised Saturnward agreed perfectly with the model. With various software patches made, the system was tested and tested again, and all the indications were that the problem was solved. Of course, it cost some of *Cassini*'s precious fuel to adjust the orbit, and a considerable amount of time and effort to de-

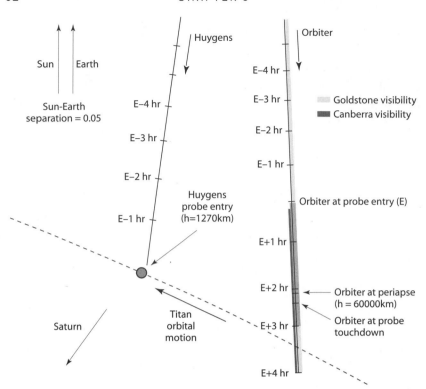

Figure 3.10.   The revised *Huygens* delivery trajectory. The orbiter and *Huygens* are on near-parallel paths, the orbiter about two hours behind and 60,000 km to one side of the probe. This kept the Doppler shift of the radio signal small enough for the weak receiver design to cope with and recover the data. (NASA/JPL)

velop the solution, to make the various software patches (which involved bringing some individuals out of retirement), and to verify that the temperatures and new longer probe mission would not introduce yet new problems. But the exercise also had the effect of focusing the teams' attention on the probe and how it would work.

## BY JOVE!

The Jupiter flyby was an important opportunity to exercise the capabilities of the teams and the spacecraft to design and execute instrument sequences, in much the same way as they were going to have to at Saturn. When *Voyager* flew by a planet in the 1980s, it would take up to six months to generate the detailed observing plan, to work out which instruments were pointed where, and when, and when the ground station would be available to receive the data. *Cassini* was to have around one hundred such encounters, so it was crucial to develop a streamlined process for generating a timeline of operations. Jupiter was the first real chance to try this out and find what had to be improved. In order to spread out the budget of a space project, the development of some capabilities (such as having the spacecraft repoint itself to make mosaics of multiple images) is deferred until after launch.

This meant that the science teams were learning as they went along, using software that was still under development, being written and improved even as it was being used for the first time—a frustrating situation! Even though the Jupiter flyby observations were officially for instrument checkout, there was, of course, the prospect of new scientific discovery, so the teams worked against the clock to get the sequences ready.

An important bonus was that the *Galileo* spacecraft, in orbit around Jupiter since 1995, was still operating. Making simultaneous measurements with two spacecraft would leverage the scientific value of each— for example, by measuring in situ the conditions of the solar wind just upstream of Jupiter, and by measuring the effects of the changing wind on Jupiter's magnetosphere.

The Jupiter encounter provided a beautiful sequence of thousands of images of the planet. These filled an important gap in Jupiter science, since the failure of *Galileo*'s antenna had meant that it did not have the capacity to send back many pictures. A novel measurement used *Cassini*'s radar to sniff out the faint microwave glow from Jupiter's synchrotron radiation belts, revealing that high-energy electrons (5–20 MeV) were more abundant than had been predicted.

Although the sequence was derailed for a few days because one of *Cassini*'s reaction wheels, used for fine pointing, seized up, overall the Jupiter encounter was a great success. The data are even now only just beginning to be analyzed in detail.

## NEARING THE TARGET

Even in October 2002, when *Cassini* was still 285 million km from Saturn—almost two astronomical units—the spacecraft's view of Saturn

Figure 3.11.    An image of Jupiter taken during *Cassini*'s flyby in December 2000. The Great
Red Spot is prominent at left; Jupiter's volcanic moon Io appears at the right and casts a neat
circular shadow on the Jupiter's cloud deck. (NASA/JPL/University of Arizona)

was an arresting sight. *Cassini*'s vantage point gave it a view never observable from Earth (but rather similar to a sketch Christiaan Huygens once
made) of the shadow of Saturn cast onto the rings, with the north pole
of Saturn just peeking above the shadow the rings made on the planet.
Titan was visible, though only as a dot a few pixels across. Things would
just get better and better from here on.

By May 2004, *Cassini* was nearing Saturn and was within 30 million
km (0.2 AU). Even from this distance, *Cassini*'s camera could see details
on Titan's surface, with twice the resolution that the HST had been
capable of a decade before. It was, however, a very different view. Instead
of being a disk, Titan was seen as a half disk. *Cassini* was approaching the
Saturnian system from a little way in front of Saturn as it was traveling in
its orbit, and so the direction of view from *Cassini* was almost at right
angles to the Sun's direction. *Cassini* took enough images to make a map
better than the HST's. These pictures required exposure times of some

Figure 3.12.   Majestic planet Saturn looms, with her rings tastefully tilted, as seen from *Cassini*'s vantage point some 285 million km (2 AU) away in October 2002, a full twenty months before arrival. In southern midsummer, the planet casts a long shadow across the rings, and vice versa. Titan is visible at the top of the image. (NASA/JPL/Southwest Research Institute)

thirty-eight seconds, but the massive *Cassini* was steady as a rock, and there was no smearing of the images. The new maps showed what was by now a familiar pattern of bright and dark regions, the strongest contrast being around the equator between long, dark branched regions and the bright Xanadu. But how should they be interpreted?

## PHOEBE

On June 11, scientists got their first taste of the kind of close-up images they hoped would soon be pouring in. *Cassini*'s arrival date had been timed so the spacecraft, while on its way in through the Saturnian system, would pass close to Phoebe, the largest of Saturn's outer moons, on June 11. Initially, the plans had been to fly by at a respectful distance of tens of thousands of kilometers, but when it emerged there was no good reason for not going closer (apart from a prudent margin of error to avoid any possible dust cloud close to the surface), *Cassini* was aimed to pass only 2,000 km away from this mysterious body—a thousand times closer than the distance from which either of the *Voyagers* had seen it.

Little Phoebe was always thought to be something of an interloper in the Saturnian system. While most of the satellites dutifully orbit in the same direction around Saturn, and very close to the ring plane, Phoebe's distant orbit goes the other way, and at quite a tilt to the ring plane, suggesting that Phoebe did not originate in the disk of dust and gas

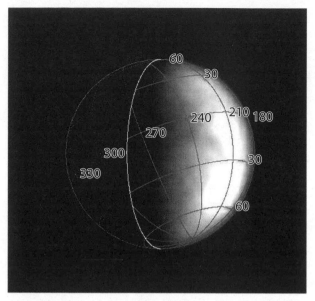

Figure 3.13.   Details started to appear on Titan from a distance of 30 million km on May 5, 2004. This image was taken by *Cassini*'s ISS camera in the same wavelength (940 nm) as the HST map in chapter 2. From *Cassini*'s viewpoint, Titan is only half illuminated. The northern high latitudes are in darkness. (NASA/JPL/Space Science Institute)

in which the other satellites were formed like a miniature solar system. How Phoebe happened to get trapped at Saturn is a question for which nobody has good answers. Textbooks often suggested Phoebe might be a trapped asteroid.

"That's no asteroid" came back the results. The reality was, as ever, much more complicated and interesting. Although Phoebe looked slightly irregular and cratered, and some largish boulders were even visible in the sharpest pictures, it was clear that Phoebe was not built like an asteroid. The staggering pictures showed bright streaks in some of the craters, suggesting that darker material had slumped away, exposing brighter ice beneath. In fact, it seemed a lot like a quiet comet!

In terms of how easy it is for things to be flung around the solar system, the ragged edge of the Saturnian system is rather nearer the Kuiper belt beyond Neptune, with its population of icy planetesimals like Pluto and its cousins, than it is to the asteroid belt. Although some scientists talked of Phoebe being a captured Kuiper belt object (KBO), the reality may be

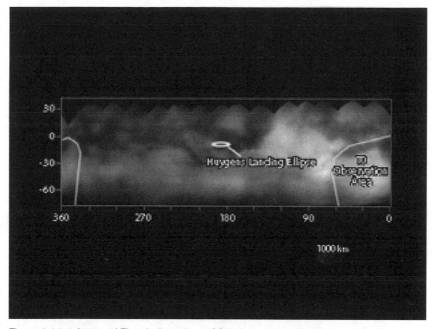

Figure 3.14. A map of Titan built up from ISS images during the approach phase, prior to Saturn Orbit Insertion (SOI). No details are visible north of about 30° north. The *Huygens* landing site is indicated to be near an equatorial bright/dark boundary, and the area to be observed during the T0 opportunity, just after SOI, is shown. The lack of details at southern midlatitudes is interesting, and is somehow intrinsic to either Titan's surface or its atmosphere at this season. (NASA/JPL/*Cassini* Imaging Team)

even more exotic. It is possible that Phoebe is the last remnant of some population of objects that is now no longer there, expelled by the gravitational turmoil of the early solar system, a unique relic preserved in a Saturnocentric orbit like a wooly mammoth in permafrost. Not a KBO, not an asteroid, but something in between.

Its composition, as measured by the VIMS instrument, included bound water, trapped $CO_2$, phyllosilicates, organics, nitriles, and cyanide compounds. The presence of all these compounds makes Phoebe one of the most compositionally diverse objects in our solar system. It also heralded the prospect that the composition of materials in the Saturnian system as a whole may be more than simple "rock and ice."

As it sped past tiny Phoebe, *Cassini* was still on a trajectory around the Sun. After this encounter in the outskirts of Saturn's extended family at a distance of 11.8 million km, it was a further nineteen days before *Cassini*

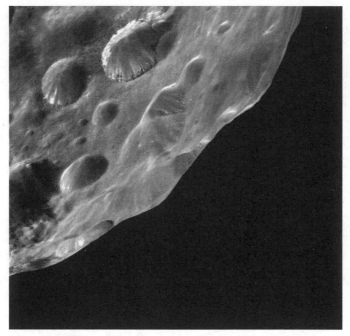

Figure 3.15.   Close-up image of Phoebe on June 11, 2004, from a distance of about 14,000 km. The surface is pockmarked with craters, but appears to have some bright streaks in crater walls, perhaps indicating the presence of ice under a coating of dark material. (NASA/JPL/ Space Science Institute)

closed in far enough for the Saturn Orbit Insertion maneuver, just 100,000 km from Saturn's cloud tops.

## IN ORBIT AT LAST

When the drama of the SOI, with which we opened this chapter, was over, there was general relief all around. Only now was it truly possible to say that *Cassini* had arrived safely. However, some days later, it emerged that not quite everything had gone according to plan. A sensitive channel on *Cassini*'s magnetometer was to have been switched on to make precision measurements of Saturn's magnetic field. It would have been a key observation because this was the closest *Cassini* would ever get to Saturn, and so the best chance of resolving fine structure in Saturn's apparently symmetric magnetic field. But among the millions of instrument commands, this one had an error: the day number was wrong by one. This sort of error causes no engineering risk, so checking in the spacecraft testbed raised no warning flags. But the scientific loss was profound.

But that small setback was just a detail in the bigger picture. And the pictures themselves were not just big—they were spectacular. This was also one of the closest encounters with the rings in the whole mission. *Cassini*'s imaging team leader, Carolyn Porco, whose own scientific interests centered on rings, was thrilled. Reacting to images showing textbook density waves—the gravitational signature of Saturn's moons in the dynamics of the ring particles—she declared them "absolutely mind blowing. Look at that. Ooh. . . . It's almost everywhere you look here, you can't miss one. They're just all over the place."

As *Cassini* shot past the rings at 15 km a second, many of the camera exposures had to be only five milliseconds long to prevent the relative motion from smearing the images. But that was plenty to reveal the splendor of the rings. Perhaps the most impressive single image was one of the Encke gap, only 300 km wide, bracketed with density waves that wrapped up at the edge of the gap to give it a beautiful, scalloped appearance.

As *Cassini* arced away from Saturn, again toward the south after SOI, Titan lumbered along above in its orbit in the ring plane, giving *Cassini* a bird's-eye view, albeit a distant one (300,000 km away), of Titan's south pole. Since this circumstance provided a good opportunity to reconnoiter Titan, the *Cassini* scientists decided to treat it as an extra, preliminary Titan flyby and named it T0.

An engine firing (a "cleanup burn") to fine-tune the trajectory after the large orbit insertion burn was scheduled for July 3, but was canceled; the arrival had been precise enough for the cleanup not to be needed. Then, for the next several days, communications with *Cassini* were difficult or impossible, as expected, because Earth and Saturn entered conjunction. This event, where Saturn and Earth are on exactly opposite sides of the Sun, occurs every twelve and one-half months. Not only is the Sun directly interposed between Earth and *Cassini* for a short while, but for several days on either side of conjunction, the Sun (which is a strong source of radio noise) lies within the beam of *Cassini*'s antenna. Making sure no dramatic events like engine firings or Titan flybys occurred during conjunctions was just one of the many rules that the *Cassini* tour designers had to follow while weaving their tapestry of orbits around Saturn.

Figure 3.16.  An image acquired just after Saturn Orbit Insertion of the Encke gap in the A
ring. Faint ringlets are seen in the gap, which proves to have a scalloped edge due to the
gravitational effects of one of Saturn's moons. The spiral structures are bending and density
waves. (NASA/JPL/Space Science Institute)

   The first orbit *Cassini* completed around Saturn was its longest—a
period of some four months—with the spacecraft still only loosely confined inside Saturn's gravity "well." Like a comet's orbit around the Sun,
this orbit was very elliptical, and without intervention would take *Cassini*
perilously close to Saturn and its rings. Something had to be done. And
so, in late August as it approached apoapsis—the farthest point in its orbit
from Saturn some 150 $R_s$ away (Saturn radii; 1 $R_s$ is about 60,000 km)—
Cassini fired its engine again for about fifty minutes to raise the periapsis
(closest approach) of its orbit to a safe altitude. The burn, followed by
some fine-tuning a couple of months later, also set *Cassini* up for its first
close encounter with Titan.

# 4. *Cassini*'s First Taste of Titan

With the tension of orbit insertion behind them, the science teams could indulge in their last speculations about Titan and get to work on the first results from *Cassini*'s initial approach and the T0 flyby. A major conference, COSPAR, took place in Paris at the end of July 2004 where these early findings were presented. Presentations from most of *Cassini*'s instruments focused on Phoebe, Saturn, and the rings, but the optical remote sensing instruments, VIMS and ISS, had dramatically new perspectives on Titan, too.

## T0: A GRANDSTAND VIEW

The ISS imager had a grandstand view of Titan's south polar regions. Not only was the resolution it could achieve rather higher than the best from Keck and HST, but these polar areas were only ever seen slantwise from Earth, at the edge of Titan's disk, where *Cassini* was looking more or less straight down on them. It would be a quantum leap forward, a factor of ten better than anything so far, and of an area never seen before.

Distinct bright and dark areas were visible, and details down to a few kilometers across could be seen—rather better than the gloomiest predictions about blurring due to the haze—although the real test of imaging detail would require getting closer than the 300,000 km distance of T0.

But these bright and dark patterns were, much like the HST map a decade before, rather inscrutable. There were some patches, some narrow

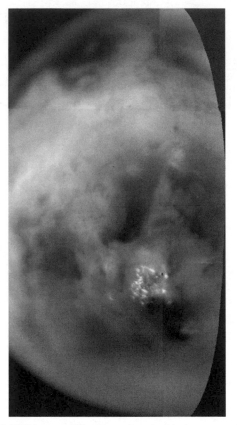

Figure 4.01.   A *Cassini* ISS image of Titan's southern hemisphere seen during the T0 flyby (see figure 3.14 for map). Intriguing dark patches and sinuous features are seen, together with brighter areas and a complex of bright clouds around the south pole. (NASA/JPL/Space Science Institute)

wiggly lines (could they be rivers?), and some rather straight-edged features, hinting at the possibility of some sort of tectonics.

And above this still-mysterious surface, the now-familiar complex of south polar clouds was still active. A sequence of images, each a few hours after the previous one, showed the clouds moving in the wind, and evolving as they did so, just like on a series of satellite pictures of Earth.

The VIMS instrument, able to probe at a longer range of wavelengths, added some "color," at least at low resolution, to our view of Titan. The VIMS team saw, much as Earth-based observers had, bright and dark

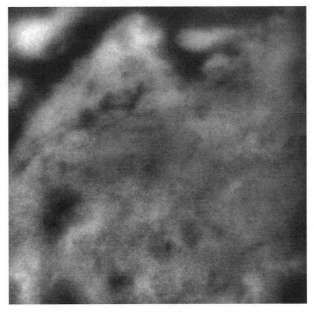

Figure 4.02.   An individual frame of the ISS T0 observation. Blurry details a kilometer or two across are visible, notably the dark "roadrunner" in the center. Geological interpretation was difficult, although imaging scientists did note the straight edge of the feature at upper left, which is suggestive of some kind of tectonics. (NASA/JPL/Space Science Institute)

areas on the surface. Some rather speculative interpretations were made of dark spots as impact craters or "palimpsests." (A palimpsest is a patch of discolored surface, the scar of an impact crater left when the combination of the softness of the crust and the length of time since the crater formed has allowed the topography of the crater to relax by viscous flow. Several such patches are seen on Ganymede and Callisto.)

Among the familiar patterns of bright and dark on the surface was the hint of some compositional differences; not all bright and dark regions were exactly as bright and dark at all wavelengths. But until the effects of different slant paths through the atmosphere and of different Sun angles could be understood, solid clues to surface composition would remain elusive.

It was easy to see in the VIMS image that Titan was much "bigger" at some wavelengths. While the near-IR light in the methane windows that allowed surface mapping penetrated (obviously) all the way to the surface and defined Titan's solid limb, at 3.3 microns, methane gas fluo-

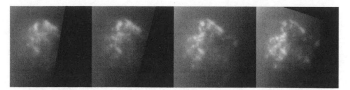

Figure 4.03.   A movie sequence of the south polar clouds. These images were taken a few hours apart, and show that the clouds around the south pole are changing on short time-scales, suggesting they might be cumulus clouds with precipitation (methane rain or hail). (NASA/JPL/Space Science Institute)

rescing in sunlight some 700 km above the surface showed that, in the stratosphere at least, the methane was fairly uniformly mixed. On the nightside, a faint glow from carbon monoxide at 4.7 microns extended some 200 km above the surface.

The VIMS results also showed clouds that could be tracked from one image to the next, indicating tropospheric winds near the south pole of only 1 or 2 m per second. Furthermore, study of the spectrum in each pixel would allow the altitude of the cloud tops to be measured, and it would later be determined that the clouds were puffing up at several meters per second, not much slower than cumulus cloud tops on Earth.

Titan was too far off for *Cassini* to use its radar instrument, but the infrared spectrometer CIRS was aimed at Titan to measure the atmospheric temperature distribution. If there were going to be any big surprises in how Titan's upper atmospheric winds change with season, they should be obvious in the temperatures too.

Within a day or so, *Cassini* had zoomed out from Saturn, leaving Titan behind for a while as the scientists puzzled over the new results and tried to anticipate what they would find later. *Cassini* reached its first apoapsis on August 27, nearly 151 Saturn radii from the planet (or over 9 million km). Like a captured comet, *Cassini* then began its long, lazy arc back toward Saturn. Near apoapsis, it fired its engine for another fifty-one minutes, bringing the periapsis up from close to the rings to some 300,000 km above Saturn. But first, its new course would encounter Titan on October 26, 2004, enabling Titan's gravity to help bring the orbit period down and to keep *Cassini* firmly in the Saturnian system far more efficiently than *Cassini*'s engine ever could.

Meanwhile, on Earth during the Northern Hemisphere summer, it was still the conference season. One engineering-focused conference was the International Planetary Probe Workshop at the end of August, held

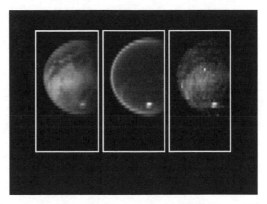

Figure 4.04.   Titan seen by the VIMS instrument during T0 at three different wavelengths. Left to right, the wavelengths are 2, 2.8, and 5 microns. The south polar clouds are bright at all wavelengths. At 2.8 microns the surface is mostly dark, with the hazy limb somewhat bright. At 5 microns, where water ice is not reflective, the correlation with the 2-micron (and 0.94-micron) bright and dark areas suggests that ice is not the main "brightening agent" on Titan. (NASA/JPL/University of Arizona)

that year at NASA's Ames Research Center. Preparations for *Huygens* were reported, notably the ongoing reevaluation of the heat shield's expected performance and the environment it would have to deal with; there were worries that the heat shield might only just be adequate. Also, the Huygens Atmospheric Science Instrument teams reported on a "dress rehearsal" of the descent, made with a mock-up of the *Huygens* probe dropped from a balloon over Sicily. Gaining familiarity with data would be key to assessing the quality of the data from Titan and developing ways to analyze the data efficiently.

But as thoughts began to turn toward the *Huygens* release at the end of the year, another planetary entry was to take place. On September 8, 2004, NASA's *Genesis* capsule returned to Earth, bearing samples of the solar wind that it had been collecting in space for three years. But instead of gliding down under a parachute to be caught gently in midair by helicopter, it smashed into the desert. The g-switches used to start the parachute deployment had been mounted the wrong way up and so had not triggered. Although most of the scientific results were salvaged, it was a very visible failure. Coupled with the loss of ESA's *Beagle 2* Mars lander nine months earlier, it was a reminder that entry, descent, and landing are always a challenge. And it made everyone nervous about *Huygens*.

Figure 4.05.  Glow-in-the-dark Titan. This presentation of high-phase (crescent) Titan images from VIMS during T0 shows, at left, a methane window image (seeing down in natural sunlight to the surface, hence the smallest-diameter image). Next is an image at 3.3 microns, where methane fluoresces. Since this glow is stimulated by sunlight, it appears also as a crescent. The third image is at 4.7 microns, where carbon monoxide causes emission in the warm stratosphere. This thermal glow can be seen even on the nightside. The rightmost image is a composite of the other three. (NASA/JPL/University of Arizona)

## TITAN IN FRONT OF THE CAMERAS

TA, the first close Titan encounter, would be the first opportunity to try out all of *Cassini*'s instruments on one of their principal targets. It was therefore a particularly busy encounter, requiring the spacecraft to pirouette and roll like Jackie Chan in a martial arts movie. In the hour or two around closest approach, the bus-sized craft would observe Titan optically, then swing around to sweep its radar beam in a raster pattern "painting" a big rectangular patch on Titan's surface, then swing through 90 degrees to aim VIMS to some well-lit regions, then swing back to point INMS forward around closest approach while doing radar imaging, tracking Titan through nearly 180 degrees as the spacecraft whipped past at 6 km/s.

As at T0, there was great media interest and JPL was abuzz with reporters and TV trucks. This was the first encounter with a new world. T0 hadn't really counted for unveiling Titan because it hadn't been close to Titan and was sufficiently long after SOI that reporters had already filed their space stories. What would *Cassini* find?

Most of the data was downlinked to the Deep Space Network's 70 m dish in Madrid, since the preferred station in Goldstone, California, which enjoys more reliable weather, was under maintenance for the latter half of 2004. The order of transmission roughly followed the order of the observations themselves.

On this occasion, the dayside part of the pass happened to be on the inbound leg, as *Cassini* was approaching Titan, and so the imaging results came down first. (Later in the mission, as *Cassini*'s orbit rotated around Saturn and so changed its orientation, the opposite would become the case.)

Thus, the early focus was on the orbiter's camera, which had already demonstrated during the initial approach and at T0 that it could see things on the surface. As images were read out and displayed, essentially in real time, scientists from all the teams gazed in wonder at the features.

The imaging team members themselves struggled to enhance their images, piece mosaics together, and write captions while the media buzzed around. At first, the patterns of bright and dark were challenging to interpret. "I just wish we knew what we're looking at," noted imaging team leader Carolyn Porco. But at least some prospects could be ruled out. "No herds of roving dinosaurs, yet."

Imaging scientist Alfred McEwen made a dramatic color composite. Typically, a color image is the synthesis of three different color channels, represented by red, green, and blue, though the wavelengths in which the three images are taken do not necessarily correspond to these colors in reality. McEwen made the blue channel in the composite an ultraviolet image taken by *Cassini*, which showed the haze in the upper atmosphere. The green channel brought out the surface features at 940 nm, while the red channel was generated from a methane band image at 889 nm, with an obvious difference between the north and south hemispheres. Even though the composite was in "false color," it was an attractive and information-rich image, and was reproduced in many magazines and presentations thereafter.

But when scientists looked in more detail at the higher-resolution images, they saw some tantalizing trends. These trends lay somewhere between the most pessimistic and optimistic predictions about blurring due to haze scattering: *Cassini*'s camera seemed to be able to detect surface features down to a resolution of about 1 km, depending on observing angle and other factors.

A very sharp and irregular boundary could be seen at the western edge of Xanadu between bright and dark terrain. In terms of its shape alone, it looked for all the world like an irregular coastline. That, and the presence of isolated irregular bright areas looking like islands, made Titan strongly reminiscent of Earth. The Aegean was a particularly appealing analogue to the region seen on TA. But were these really coastlines?

Also visible were some highly intriguing dark linear features embedded in bright regions, sometimes branching. They looked as if they might

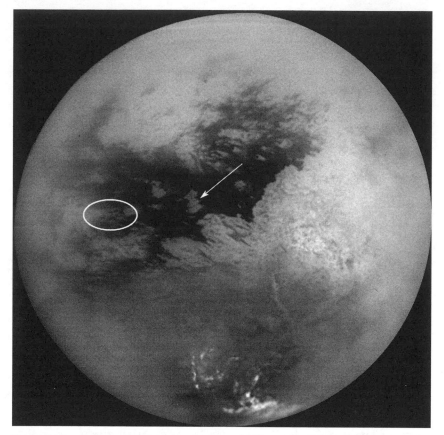

Figure 4.06.   A contrast-stretched and enhanced ISS mosaic of Titan at 940 nm from TA images. The south polar clouds are visible at the bottom. Xanadu is at right, the dark area in the center is Shangri-La, with the bright feature Shikoku arrowed. The ellipse shows the expected *Huygens* landing site. (Adapted from NASA/JPL/University of Arizona image)

be river channels, but in at least one case, the dark line went all the way across the bright region. (This area was informally referred to as Great Britain, since that's what it looked like, but its official name became Shikoku.) Such behavior was not to be expected from a river, which flows from the middle of a continent, where it rises, to the sea and not from one sea to another. So the interpretation remained tentative.

But it was puzzling. The morphology of the bright and dark regions was similar to a coastline, but if the dark regions were liquid, there should have been a specular glint—a mirrorlike reflection from the smooth surface of a hydrocarbon sea. And none was seen, either by *Cassini*'s camera or by ground-based telescopes such as the Keck. Scientists struggled to

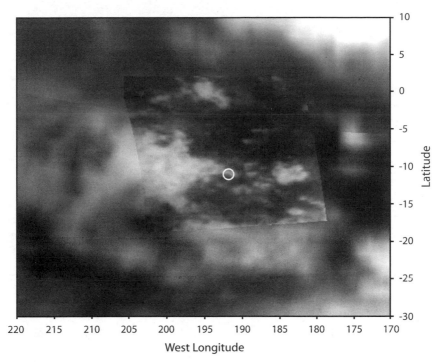

Figure 4.07.   A mosaic of VIMS data showing the *Huygens* landing site region, the inset area being at higher resolution. The small circle denotes the best a priori estimate of the landing site, at the edge of the bright feature. (NASA/JPL/University of Arizona/US Geological Survey)

reconcile the observations with their expectations, especially as specular glints had apparently been seen in the 13-cm Arecibo radar observations. The lack of an optical specular glint did not accord with that observation. Among the ideas thrown around were fluffy aerosols floating on the liquid, making it rough at micron wavelengths but smooth at centimeter wavelengths. However, this was clutching at straws. In principle, such aerosols should sink unless they trapped air to remain buoyant because the density of liquid ethane is quite low. Another possibility was that the surface was too rough for an optical glint—perhaps due to wind-driven surface waves. Or, perhaps, there was no liquid after all and we were being fooled by our hopes.

It became clear from the TA images that the probe was destined to descend over an interesting area. The target region was at the boundary between bright and dark terrain. With luck we would get to see both close-up and learn what bright and dark really meant.

## A DIRTY EXOSPHERE

As *Cassini* flew above Titan during TA, a small aperture pointed in the forward direction (the "ram direction"). This aperture guided gas molecules into a mass spectrometer—the ion and neutral mass spectrometer (INMS). This instrument counted the molecules, and as *Cassini* swung down to 1,200 km altitude and back up, it yielded profiles of the abundances of different compounds. In fact, at these very high altitudes, lightweight methane is more abundant than molecular nitrogen, which has a much larger molecular mass.

The INMS had been used during SOI to sniff the "atmosphere" sputtered off the rings, but TA was its real debut, and some software problems in the instrument had only just been ironed out. But this first Titan encounter worked well. Remarkably, the INMS did not just detect a handful of molecules like $N_2$, $CH_4$, $C_2H_6$, $C_2H_2$ and so on (with molecular masses of 30 or less), but rather yielded a complex mass spectrum of compounds with molecular masses up to 100 and more. The instrument could only go as high as mass 100, but even at 100, there was plenty of material. No one expected such a zoo of large molecules at this altitude.

The INMS data were crucial in evaluating how low *Cassini* might go in the future. Indeed, it was so much so that these data were labeled "critical" and were downlinked especially early. If the density of the atmosphere at high altitude were greater than expected, then the attitude control system would need to fire its thrusters more frequently to keep on track. In the extreme case, the drag torques, which depend on the orientation of the spacecraft in space, would be too large for the spacecraft to deal with, and flybys would have to be made at higher altitudes than originally planned. This would essentially require a slight redesign to the whole tour.

And at this time, all the worry about the *Huygens* probe was still present. Although the INMS data pertained to altitudes far higher than those where peak heating would occur, if the INMS results were really surprising, doubts would be raised. Despite the surprising abundance of heavy molecules, the atmosphere overall seemed to be within the range of uncertainties allowed by the models that had been used. But atmospheres can be fickle things, and no one would be comfortable until more data had been obtained; the *Mars Global Surveyor* spacecraft nearly broke off

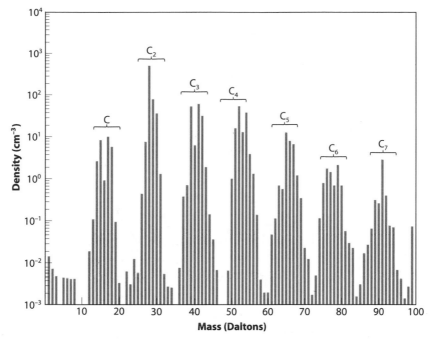

Figure 4.08.  A mass spectrum of Titan's upper atmosphere recorded by the INMS instrument above 1,100 km altitude. Heavier molecules—remarkably heavier than C6 compounds like benzene—are toward the right and were surprisingly abundant. (NASA/JPL/University of Michigan)

one of its solar panels when the density of the Martian atmosphere suddenly increased in response to a dust storm.

## RADAR IMAGE

As *Cassini* returned to the surface, although the ISS results were impressive, it was no secret that *Cassini* had a radar for the express purpose of seeing Titan through the haze. On this first close flyby, it would make its first observation of the surface, and that would set the stage for the rest of the mission. Although it seemed highly improbable, nobody could be sure that the 1990 prediction made by a prominent radar astronomer who wasn't on the team—that Titan was featureless to radar, an uninterpretable mess of subsurface reflections—might not turn out to be true.

The radar team would not get its data until much later, after media prime time. It was ensconced in a design lab run by Steve Wall, the deputy team leader (who, like almost everyone working on *Cassini*, works on other projects too), where computer projectors could show large images on three of the four walls of the room. This was to become the "radar war room" for a few days a year. Not only was it an ideal venue for studying the new data, but apart from a visit by a *60 Minutes* film crew, it was secluded enough to allow the team to be productive.

The RADAR team gathered early in the morning on Wednesday, October 27. Even before the science team in the war room saw a huge image of Titan for the first time, a few engineers in an anonymous building a quarter of a mile away got a sneak peek as they massaged the raw data into an image. By the magic of synthetic aperture radar (SAR) processing, the bits of Titan contributing to the echo could be teased apart, making an image with pixels far smaller than the width of the beam itself. Processing the raw radar data from each of the five overlapping beams required several hundred sophisticated calculations for each burst of radar pulses. It took around an hour on a fast PC to work through the whole image with its thousands of bursts. And unlike most of the other instruments, which had had plenty of opportunity to refine their procedures and methods by looking at Saturn or other targets in the months and years up to this point, this was the first time ever that the mapping mode of the radar was being used.

The very first image looked pretty ratty, striped with dark lanes where the beams overlapped. Because each of the beams hit the surface at a different angle and from a different distance, the signal strengths were different, and so it was difficult to get a uniform brightness across the scene unless all the parameters were set correctly. Some of these parameters depended on the exact trajectory. As it turned out, *Cassini* actually flew past Titan on this first flyby 26 km lower than was planned for, the difference being as much due to Titan's position not being known precisely as anything else, but this would not be measured until some hours later. The team tweaked the parameters in the processor and waited again for the sorcery of fast Fourier transforms, pulse decompression, and correlation to work their magic.

"We own the mission," observed the usually reserved veteran radar astronomer Steve Ostro. Even the first bad SAR processor run had shown that Titan's surface was rich in features. There was definitely stuff down there, though it was too early to make sense of it. When the next run came through, the dark gaps between the beams were full of detail too.

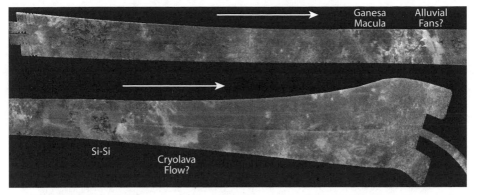

Figure 4.09. The long swath of radar data acquired during TA in October 2004 showing a bewildering array of unfamiliar features. The swath is about 180 km wide at its narrowest point and is so long it is cut in half to fit on the page. The circular feature Ganesa Macula has had the most attention so far, being a possible cryovolcano. Bright triangular features may be alluvial fans. A bright lobate feature two-thirds along the swath could be a cryolava flow. The dark spots like Si-Si might be lakes, but evidence at this point was not compelling.

A *60 Minutes* presenter asked Larry Soderblom if we knew enough to be surprised. "Oh, we knew we'd be surprised . . .", and sensing a certain unflattering probing in the question, added for good measure "so we were right!" Titan's surface seen at radar high resolution certainly didn't look much like any planetary surface anyone was familiar with. It didn't even seem to resemble the parts of Titan that had just been seen optically. At closest approach when the radar was operating, *Cassini* was sweeping over fairly high latitudes, so the coverage did not overlap areas that had been seen by ISS, either earlier on during the TA encounter or before. It was terra incognita. But even allowing for the different location, there was really nothing in the radar image that could be compared with what was seen optically.

The sharpest part of the image was dominated by a large, dark circular feature, 180 km across, with a bright edge. For an hour or two it seemed to be a crater, but like some Rorschach test that probes subjects' preconceptions more than their vision, that impression didn't last. Somehow, magically, the circular feature popped itself forward, becoming a dome instead of a crater. And the picture made more sense. As well as some of the chaotic background looking like small lava flows, the bright wiggly lines inside the dome seemed to fit as small canyons flowing down the sides of the dome, broadening as they went. And it looked strikingly like the volcanic features called pancake domes on Venus. For the moment, its identification as a volcano is tentative, and the structure bears the name "Ganesa Macula."

Lots of other features were there too: a region of giant bright fingers, looking like the lobes of some great lava flow; dark crescentlike and angular patches; and some bright, striated triangular areas that seemed to connect to bright, sinuous lines. Perhaps these were alluvial fans, sheets of debris deposited at the mouths of canyons by flash floods. But that was an entirely speculative suggestion at the time. We still didn't really know what we were seeing on Titan.

The RADAR team leader, Charles Elachi, who had a couple of years before become the director of the Jet Propulsion Laboratory, drew particular attention to some irregular dark regions in the image. Some of these formed a vaguely connected archipelago with a couple of angular edges toward the top. "Si-Si the Halloween Cat," Elachi called it, suggesting maybe they could be lakes of liquid hydrocarbons. The interpretation was a little speculative and the name was clearly not an official one, but Elachi's sense of the moment—it was three days before Halloween— probably got some Titan results into newspapers that wouldn't otherwise have carried them.

Another, more obscure, feature of the radar data was also presented at the press conference that week (by the first author of this book, although the bulk of the work had been done by Mike Janssen and others on the team). This was microwave radiometry—using the radar receiver to detect the faint natural radio glow from surfaces rather than radar echoes. If you know something about the temperature of the surface and can measure its microwave brightness (usually expressed as a "brightness temperature"), then you can deduce how good a microwave emitter or reflector it is, which may help to rule out or favor particular kinds of surfaces. Those covered in organics, for example, would be good emitters and would have high brightness temperatures, whereas dense rock or metal (or liquid water, for that matter—not that water was expected in this case) is a good reflector and so would basically reflect the cold of space without adding much emission of its own.

The brightness temperatures were quite high at 70–80 K, suggesting on balance that Titan was quite a strong emitter. Titan was definitely no iceball on the outside: it had to be largely covered in organic material. The change of brightness temperature with view angle also suggested the same interpretation. The full story of Titan's surface composition would take some time to emerge—indeed, it is still struggling to do so as we write—but Titan was clearly an exotic, organic-rich place even on the basis of those early observations.

The radar also swept a region in a low-resolution "scatterometry" mode, in which it just measures the radar reflectivity in the beam without the clever SAR processing. This showed an impressive difference between Xanadu and the darker region to the west that the imaging team had observed: Xanadu was considerably more radar-bright, perhaps meaning rougher terrain. The scatterometer also hit the probe landing site, in an attempt to guess whether the probe was in for a splash or a crash. But it was known from the optical images that the landing site was in a rather mixed-up area with some bright and dark terrain, so the low-resolution scatterometry and radiometry couldn't tell us much yet.

A final piece of data was a profile of altitudes from the radar as it looked straight down on Titan. The distance from *Cassini* to the surface could be measured with an impressive precision of around 50 m, but it seemed that, even over several hundreds of kilometers, there was hardly any relief on Titan, maybe only 100–150 m of elevation change. This meant that the slopes on the scale of tens of kilometers were very shallow—only a fraction of a degree, like the northern plains on Mars, sedimentary basins on Earth, or the flat terrain of Europa.

## A HURRIED REPLANNING

In addition to the frenzied activity associated with the flyby itself, some project business also needed to be attended to. In particular, true to the adage that "No battle plan ever survives contact with the enemy," a new compilation of data on Iapetus had introduced uncertainties about exactly how massive the satellite Iapetus was. Formally, its density was now estimated at 1.25 + 0.17 times that of water, which was barely consistent with the previous estimate of 1.02 + 0.1. Normally, this would not have been a problem, but after the trajectory redesign to accommodate the *Huygens* probe receiver problem, both the probe and the orbiter would independently fly rather close to Iapetus, arcing out before swinging back in toward Titan for the encounter in January. Because the heat shield design was perceived as close to the limits of what could be tolerated, a

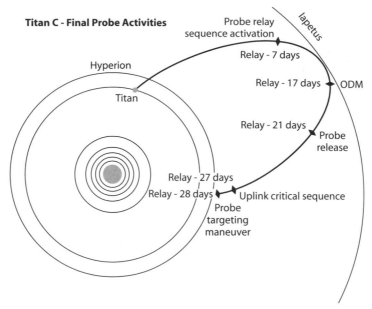

Figure 4.10.  *Cassini*'s third orbit around Saturn, setting up the delivery of the *Huygens* probe. In contrast to the geometry suggested in the artist's impressions, *Huygens* was released while *Cassini* was flying away from Titan, not toward it. In fact, both probe and orbiter flew rather close to Iapetus before arcing back toward Titan. (NASA/JPL)

change in the angle at which the probe entered the atmosphere could cause problems.

And if Iapetus's mass was uncertain, then the amount by which it would tweak the trajectory, and thus the impact point and entry angle of the probe, would also be uncertain. NASA was not in a mood to tolerate uncertainty, especially after the *Genesis* failure, so the trajectory had to be changed.

This was bad news and enormously frustrating because it was one of those things that would probably just go away with a bit more data. It was late in the day for mission changes to be made, and the navigation team was working hard. The trajectory could be adjusted to give Iapetus a wider berth, to be sure, but not without a cost. First, the changes would expend precious fuel, and by changing the flyby distance from 55,000 km to 120,000 km, the resolution of the Iapetus images would be reduced by a factor of two. Second, any changes to the trajectory or timing meant that the science teams would have to redesign all the observations in the second and third flybys at the same time as designing the observations for later in the tour and analyzing the data from the first encounters.

Third, whether redesigned or not, some observations were highly sensitive to small changes and were wrecked.

The TB encounter was to have been high enough (2,200 km) that *Cassini* would be well above Titan's atmosphere, and so it could use its reaction wheels for fine-pointing to make excellent optical images. The new trajectory brought the spacecraft down to 1,200 km, so low that it would have to use its thrusters, which would make for jerkier motion. Though the images would nonetheless be of remarkable precision, the jerkiness would be enough to make mosaic images not match up and introduce some blurring.

The T3 encounter featured a radio occultation, where Titan would pass between *Cassini* and Earth, and the atmosphere would be profiled by radio signals at several wavelengths. These would be best if the spacecraft were close to Titan at the time, so that the beam would not have spread out much, and if the spacecraft's position as seen from Earth moved slowly relative to Titan. The new T3, at a height of 1,580 km instead of 1,000 km, was going to be far poorer than had been planned. And so a debate ensued as to whether it made sense to do a bad occultation in T3 or scramble to do something else. Sadly, it was too late to make any major changes to TB.

A trade-off was worked out. The radar team would observe on T3 instead in exchange for giving up their observations on T12, which had a geometry with a good radio occultation. Although there was some initial reluctance to the trade-off, it worked out better for everyone. The radar team was particularly pleased, since T3 in February 2005 would come in what would otherwise have been a long wait between TA and the next radar imaging during T7, almost a year later.


## DIGESTING THE DATA, AND A NEW ENCOUNTER


After dumping its data from TA, *Cassini* flew by Saturn and also started to snap some of the other satellites. It got a long-distance look at Tethys, showing its old, cratered surface with much better clarity than *Voyager* had done. The second orbit around Saturn was only half as large as the first and so was faster. The second Titan encounter (TB) was on December 13, 2004.

Figure 4.11.  Two ISS views (from late 2004 and early 2005) of Titan's haze, which turned
out to have a much more complicated vertical structure than had been anticipated, especially
near the north pole (in darkness at top of right panel). (NASA/JPL/Space Science Institute)

TB was more or less over the same region of Titan as TA, but was
valuable even so. Seeing the same places at slightly different angles is
important for understanding the reflection properties of surfaces. Often,
what can look like a change in surface coating is just an effect of different
viewing geometry. A puddle can look black or like a mirror. Also, even
if the geometry were exactly the same, there isn't enough time in a flyby
to point the instruments everywhere that can be seen, nor is there enough
room on *Cassini*'s data recorders to store all the data that could be taken
if there were. So TB, especially since it had been somewhat compromised
by the Iapetus trajectory change, was not much of a headline grabber but
was rather an opportunity to fill in and build up the accumulating mountain of data. Close-in views of the haze showed it to be rather more
complicated than was previously thought. And one big surprise was that
the clouds that had been so prominent around the south pole in T0 had
essentially disappeared.

Ground-based monitoring by graduate student Emily Schaller and her
colleagues at Caltech, using an impressive combination of a fourteen-inch
telescope in New Mexico and the 10-m Keck telescopes in Hawaii,
showed that Titan's clouds came and went. Perversely, although there
were prominent clouds in Keck images on October 7 and November 4,
whose presence was also indicated by deviations in Titan's lightcurve of

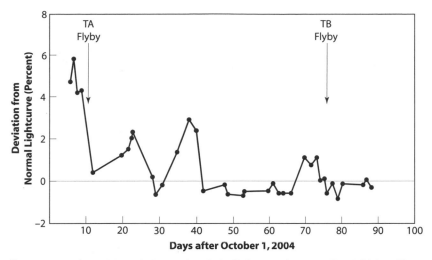

Figure 4.12. Ground-based observations help fill the gaps between *Cassini* flybys. These data, from Emily Schaller at Caltech using a small telescope, show how the cloud-induced brightness declined abruptly after TA. By TB most of the clouds had disappeared.

several percent measured with the "amateur" telescope, the clouds seemed to disappear almost entirely, with only a 1 percent deviation indicated, during the TA and TB encounters. Titan's clouds seemed almost to be deliberately hiding from *Cassini*. In reality, of course, the weather patterns were just changing as Titan's seasons wore on, and the clouds were just the sputtering remnants of southern summer storms.

One observation on TB was important enough to the project for it to be given a special "protected" status. That meant being stored in a special partition on the data recorders such that it would not be overwritten until ground controllers were sure the data had been safely received on the ground. (The data from the *Huygens* probe, radar observations of the landing site, and some other special observations were accorded similar status.) This observation was the occultation of two stars, Spica and Lambda Scorpii (also known as Shaula, the twenty-first brightest star in the sky), to be observed by the ultraviolet imaging spectrometer (UVIS). These observations would record the abundance of methane and other gases at high altitudes, and would be important in deciding how low *Cassini* would be able to go in future flybys.

This UVIS observation and the INMS data from TA encountered different bits of the atmosphere (different latitudes, different times of day, and so on) but seemed broadly concordant. But the Titan Atmosphere

Working Group (TAMWG)—projects have an impressive proliferation of teams, groups, committees, and so on, which belies the fact that basically the same few scientists are participating in them all—had discovered an unsettling discrepancy. The engineers monitoring the *Cassini* attitude control system were able to estimate the atmospheric density too, by gauging how much fuel the spacecraft had to use to counteract the torque on its appendages. Their estimate was four times higher than the INMS one, which meant the density might be too high to allow *Cassini* to make the flybys planned for 950 km up.

Some INMS scientists were inclined to dismiss the indirect engineering measurement; after all, INMS had been designed to precisely measure Titan's atmospheric density. How could a bunch of engineers armed only with an angular momentum estimator and a thruster duty cycle plot be telling them they were so wrong? Probably someone's assumptions were wrong somewhere. But in the end, it wasn't what INMS measured that would determine the flyby altitudes, whether that was the real density or not, but what the attitude system could cope with.

Much discussion and investigation failed to resolve the discrepancy, and some months later, it would be decided that the flyby altitudes had to be increased somewhat, just to be safe. Yet more replanning of observations would be needed.

## THE "SNAIL"

The weeks following TA gave the teams time to start making interpretations of their data. One intriguing observation came from the VIMS instrument—in fact, from a rather unpromising moment. Near to Titan, *Cassini* needs to use its thrusters to turn quickly, while farther away, well above the atmosphere and where turns to track a target like Titan can be slower, it can use reaction wheels. But for the attitude control system to make the transition between these very different systems takes twenty minutes or so, during which time *Cassini* can't make any exotic turns, although it can track a specific target. During one of these transitions, the VIMS instrument stared at a random spot on Titan. The spectral image cube it generated—only sixty-four pixels square, but with many hundreds of wavelengths—was impressive.

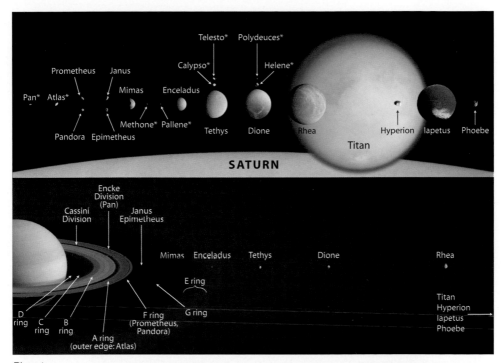

Plate 1.a

Plate 1.b

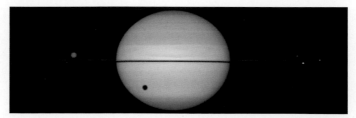

Plate 2.a

Plate 2.b

Plate 2.c

Plate 3.a

Plate 3.b

Plate 4.a

Plate 4.b

Plate 5.a

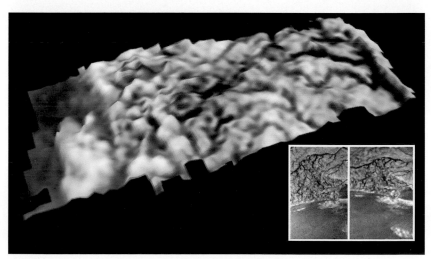

Plate 5.b

Plate 6.a

Plate 6.b

Plate 6.c

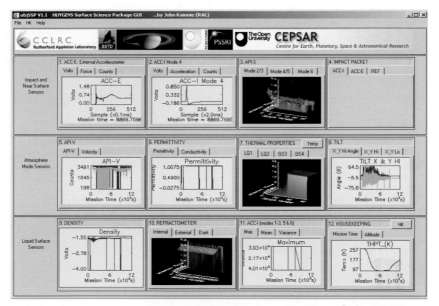

Plate 7.a

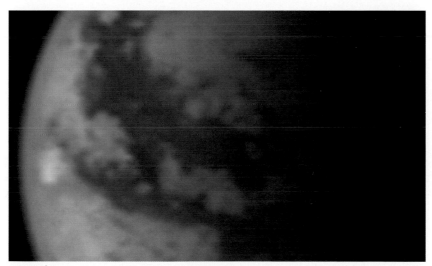

Plate 7.b

Plate 8.a

Plate 8.b

Plate 8.c

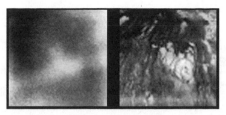

Figure 4.13.   What a difference a micron makes. Both of these VIMS images are about 100 km across. The one at left is at 0.94 microns, the same as the ISS camera, and the right frame is at 2 microns, where the haze is much thinner and reveals much more surface detail. Some suggestion of flow features is seen at right, and the spiral was interpreted by some to be a cryovolcanic structure. (NASA/JPL/University of Arizona)

First of all, it showed how much clearer the atmosphere was at a wavelength of two microns rather than at one micron for showing small-scale features. A number of irregular features looked remarkably like river valleys. And another feature looked like some sort of coiled spiral, which was nicknamed "The Snail" (and later officially named Tortola Facula). It was interpreted by some members of the VIMS team as a volcanic structure. More skeptical members of the planetary community believed that the interpretation was rather speculative, and a more dismissive nickname, "The Cat Poo," was advanced. Whatever it was, everyone agreed it was an interesting structure.

## CHRISTMAS EVE: PROBE RELEASE

It was going to be a busy New Year, but a dedicated few on the project attended a major event over Christmas too: the release of the *Huygens* probe on Christmas Eve (U.S. time) 2004. The command to release the probe was sent. At the specified moment, a pulse of electricity fired the pyrotechnics that released the probe. Three large springs, held compressed for the eight years since the probe was mated onto the orbiter, pushed the probe away. *Huygens* was guided by rollers along three spiral tracks, with the intention of giving it a speed of about 30 cm/s and a rotation rate of 7 rpm. As the probe separated, its umbilical stretched taut and yanked out the special low-force connector that was its only remaining link with *Cassini*. The probe was on its own.

On the ground, controllers could verify that the separation event had gone as planned. The magnetometer on *Cassini* saw a faint,

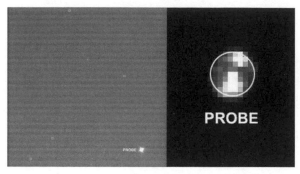

Figure 4.14. Have you seen this space probe? *Cassini*'s ISS snapped this picture showing the *Huygens* probe against the background of stars on December 26, 2004, two days after the probe separated from *Cassini*. The blowup (*right*) shows some shadow detail on the probe, 2.7 m across but some 52 km away. (NASA/JPL)

spin-modulated signal begin and decay as the probe spun off into the distance. Also, just as the springs spun up the probe, they kicked back on the orbiter, giving the massive spacecraft a small spin impulse in the other direction. The situation was complicated slightly by all the fuel sloshing around in *Cassini*'s now half-empty tanks. After a few minutes, the attitude control system kicked in and stopped the rotation induced by the separation. The reaction on the orbiter was perfectly consistent with what had been expected. So far, so good.

*Huygens* was now well and truly on its own. It was set to coast, asleep and gently spinning, for twenty-two days, during which time Titan would sweep around Saturn one and one-half times before probe and moon would meet at the appointed time on January 14, 2005. Before then, it was going to get colder than at any other time during its mission. All its systems were shut down, except for three quartz clocks, set to wake it up a few hours before it entered the top of Titan's atmosphere. To stop it from getting too cold for crucial components like the batteries to survive, it was kept warm with a couple of dozen pellets of plutonium, which supplied just enough heat to enable *Huygens* to be safely resuscitated. In preparation for *Huygens*'s descent, *Cassini* would reorient itself and make another burn of its engines. This burn would slow *Cassini* down so that it would arrive near to Titan several hours after *Huygens* and could relay the radio signals from *Huygens*.

Within a few hours of the separation, *Cassini* took a mosaic of images using its wide-angle camera to locate the probe. One of the images showed *Huygens* as a small circle. Once the probe was found, which had to be done quickly in order to update the command sequence on the spacecraft for taking the subsequent images, a new mosaic was made using the narrow-angle camera. *Huygens* was by now many tens of kilometers away. It was an impressive feat, and there had been many memos and discussions over the years about whether implementing such an observation was worthwhile and how it should be done. *Huygens* was apparently pointing in the right direction, and reassuringly, there was only one *Huygens* rather than a constellation of bits. (In August 2002, the Spacewatch telescope run by the University of Arizona had observed three fragments moving in space at roughly the speed that the *Contour* spacecraft should have been traveling. Apparently, it had broken up or exploded during the firing of its solid rocket motor to depart from Earth.) The position of *Huygens* on the sky would also add confidence to the navigation solution of where the probe would arrive at Titan. Reassuring as the picture of *Huygens* was, a little circle in the blackness of space, many couldn't help thinking that it was just such a picture that had been the last anyone saw of the *Beagle 2* Mars probe almost exactly a year before.

## A WALNUT ON NEW YEAR'S DAY: THE IAPETUS FLYBY

When *Huygens* was released, both *Cassini* and the probe were still arcing away from Saturn before they swung back toward Titan and the next Saturn periapsis. They arced out almost as far as the orbit of Iapetus, and as luck would have it, Iapetus happened to be in the vicinity (which led to the TB/T3 redesign discussed earlier). As the *Huygens* probe drifted silently past Iapetus, its three clocks ticking off the seconds until its moment of truth at Titan, the *Cassini* orbiter was busy.

The camera team even made some shots in the dark—literally. Some regions of Iapetus were in the line of sight from *Cassini*, but were on the nightside. For a normal camera that might be problematic, but *Cassini*'s rock-steady pointing performance meant that the camera could dwell for a long exposure while tracking Iapetus—long enough that the hidden area could be seen by light reflected from Saturn. Just as the night side of the moon can sometimes be visible as Earthshine, so too was Iapetus brought out of the shadows. The image taken was a striking one, showing

Figure 4.15.    Iapetus. *Left*, an ISS image of Iapetus's nightside, lit by faint Saturnshine. The trailed stars in the background indicate both the relative motion of the orbiter and Iapetus during the eighty-two-second exposure and how well it has been compensated by the space-craft pointing. *Right*, a more conventional sunlit image taken December 31, 2004, some 172,000 km away, showing the terminator at top and the bright/dark boundary just below. An amazing surprise was the 13 km—high band running around Iapetus's equator, visible at the bottom of the image. (NASA/JPL/Space Science Institute)

not only otherwise-hidden features on Iapetus, but also long star trails, just like a long-exposure photograph of the Earth's sky.

The new images laid out the puzzling distribution of dark material that gives Iapetus its yin-yang appearance, like a tennis ball with one of the two "halves" painted black. Moreover, the color filters on *Cassini*'s camera made it much easier to discriminate which areas on Iapetus were dark because they were covered in dark brown material, and which looked dark because they were in shadow.

One early result from the images was that it could be seen that the dark material had been dusted onto Iapetus's surface—either sprinkled in from elsewhere in the Saturnian system or, conceivably, sprayed out from some volcano on Iapetus (although none were identified). It had not, as the much poorer *Voyager* images had allowed, flowed on the surface.

The dark material aside (whatever it is—VIMS data are providing some clues), it was clear that Iapetus's surface was old. It had many large impact craters, the scars of a heavy battering early in its history. The largest of these was almost one-third of the diameter of the body itself. This is, in some ways, not surprising. Many worlds, Mimas and Tethys among them, have such a crater. The reasoning is as follows. There are many small craters and few large ones, and large ones tend to be easy to spot. But no crater can be much larger than one-third of the planetary diameter, or the world would have been blown apart and we wouldn't be able to observe it. So, statistically, one doesn't find more than one or two "third-of-a-world" craters, because they belong to the class of planet-busting impacts. One of the only indications of surface changes on Iapetus was a rather impressive landslide deposit, where the walls of one of the craters had slumped.

But one feature of Iapetus really stuck out—literally. A prominent ridge ran around its equator, as if the world had been made in two halves and the glue had oozed out at the seam. Iapetus looked like a walnut. Theoretical work in the following year would determine that Iapetus's strange shape had rather important implications about when it—and thus the Saturnian system as a whole—formed.

## ENTRY OBSERVATION CAMPAIGN

The end of the year marked not only the Sun reaching its highest southern latitudes on Earth, but also the closest approach to Earth Saturn had made for many months, an ideal opportunity to observe it. And this time there was an added motivation: *Huygens*'s imminent arrival meant there would be ground truth to any observations. Knowing what the conditions were at the landing site from the probe measurements would solve many of the unknowns in interpreting the ground-based results.

Conversely, ground observations would give context to the *Huygens* results. The experience of encountering Jupiter with the *Galileo* probe in 1995 underscored this point. The probe had indicated surprisingly dry conditions, which were difficult to reconcile with previous measurements and theories of how much water should be present on Jupiter. The answer was that the probe had just happened to descend in a region of downwelling atmosphere, which had been dried out. On Earth, such downwelling masses of dry air are a characteristic of the Hadley circulation: warm air rises near the equator, its water falls out as tropical thunderstorms, and the dry air descends at latitudes around thirty degrees from the equator, which is why most deserts on Earth are at these latitudes. On turbulent, fast-rotating Jupiter, there is not the same large-scale circulation, and these downwellings occur in much more localized swirls and bands. It turns out that such downwelling regions occupy only about 1 percent of Jupiter's area, and so the probe was very unlucky (or lucky, depending on your point of view) to encounter one. But no one would ever have known that the probe's environs were special, had it not been for the rather unglamorous observations from the ground. In particular, Glenn Orton at the Jet Propulsion Laboratory had monitored Jupiter's appearance at five microns and saw that the probe had entered a region particularly bright at this wavelength—"a five-micron hot spot." The reason it was bright is not that it was warm, but rather that the downwelling dry current was fairly free of clouds, and so the atmosphere was clear. This allowed the five-micron glow of Jupiter's hot interior, normally blocked by clouds, to blaze through, giving the hot appearance. It was an important lesson. Sending a single probe to a planet is a risky proposition both scientifically as well as in the engineering sense, and simultaneous observations from the ground can be critical in interpreting a probe's data.

And so there was a concerted effort to observe Titan at and around the time of the probe encounter. Telescopes on Hawaii and in Chile attempted to measure Titan's winds by the Doppler effect on narrow spectral lines. The adaptive optics imagers were busy searching for cloud activity. With *Cassini* data coming in, nobody cared about their surface maps anymore, exciting as they had been a year before.

It might be thought that all this activity was superfluous, given that *Cassini* was right there at Saturn. However, even *Cassini*'s spaceworthy complement of instruments launched in 1997 couldn't match all the heavy and capable instruments that had been fine-tuned on Earth for nearly a decade since, such as the spectrometers for measuring winds. And more important, *Cassini* had to spend a lot of its time pointing its high-gain antenna either at Titan (to receive the probe signal) or at the Earth, to relay the data to the ground. So during these times, optical observations of Titan were impossible. An added complication was that activity, and especially data recording, was restricted during the "critical sequence" to support the probe data reception and downlink. So *Cassini*'s metaphorical hands were tied, just to be safe.

One other observation was to be attempted. It was a bit of a long shot—namely, to observe the "meteor trail" made by *Huygens* in Titan's atmosphere. After all, *Huygens* was a pretty good-sized meteor. Also, there is a peculiar property of atmospheres with carbon and nitrogen (C and N) in them: when heated or shocked, they emit a brilliant violet ("CN") glow. In fact, there was concern, among all the other phantoms of risk being chased by the various reviews of the mission, that this violet light might penetrate the heat shield. Instead of being absorbed at the surface of the shield, like the direct aerodynamic heating, where it would be safely insulated from the probe structure, this violent violet radiation might warm up the bondline where the adhesive held the ceramic tiles onto the structure. If that were to happen and the adhesive were weakened, tiles might fall off and the shield would burn through. The loss of the *Columbia* space shuttle in early 2003 due to a hole in its heat-resistant but brittle wing structure had made the entry aerothermodynamics community particularly alert to such things. A special test was arranged at NASA's Ames Research Center with a powerful xenon lamp to check that it was opaque to violet light. The test posed some administrative problems: French industry was understandably reluctant to have foreign competitors analyzing its proprietary materials, used in ballistic missiles. But in the end, the test confirmed as expected (the stuff was brown after all, so one could tell by looking that it probably absorbed violet light pretty well) that there should be no risk.

....................................................................................................

## RALPH'S LOG, 2001–3

Would we be able to see it? Would the probe's fiery entry be observable from Earth? I can't remember how many times I've been asked this question, but it is more than once, and out of habit I would dismiss it. The probe is tiny, and Titan is so very far away. Not a chance.

I started to think about the problem in more detail at a *Huygens* meeting in ESTEC in late 2001. I remember when I was a young trainee engineer at ESTEC ten years before, in some idle moment (I think inspired by Hubble Space Telescope images showing a storm on Saturn) Jean-Pierre Lebreton suggested I should look into what would happen if an asteroid or comet fell into Saturn. At the time, I knew nothing about space impacts and next to nothing about planetary atmospheres. I did know that such a thing sounded pretty improbable, and I had lots of other things to do, so didn't explore the idea very much. Naturally I was kicking myself three years later when a whole fusillade of cometary fragments plunged spectacularly into Jupiter.

And so, when Jean-Pierre mused about detecting the *Huygens* entry, I gave it a little more thought. The strange chemistry of the Titan atmosphere is such that there could be some strong optical emission from the shockwave—emission in a narrow bandwidth that might make the event detectable. After all, aurorae are faint and difficult to see except on dark nights on Earth, and yet we can see aurorae on Jupiter and Saturn with the Hubble Space Telescope. Maybe *Huygens* would be visible too.

When one gets the idea that something is detectable, it is a potential discovery, and in some circumstances it might be advantageous to keep it quiet. However, I'm not really an observational astronomer. Others would be better placed to get telescope time to observe the event, and might think of better ways to observe it than I would—better techniques, better filters and instruments. And apart from anything else, on the morning of January 15, 2005, I was expecting and hoping to be busy, awaiting the data transmitted back from the probe itself. So it made sense just to advertise the possibility in the hope of inspiring someone to do it. And a publication is a publication after all.

I chose the journal *Astronomy and Geophysics*. They had a reasonably "popular" bent, a short publication cycle, and most important, published papers with attractive color figures. This avenue would be ideal—a nice marriage of astronomy and physics, and an attractive forum to build interest in *Huygens* and Titan. A journal article could easily be distributed to the observing community, giving them the scientific and technical case to justify their proposals. The article was printed in 2002, together with an eye-catching artist's impression of *Huygens* with nice violet CN emission around its heat shield by my friend James Garry. One of his favorite aphorisms is that a technically flawed but well-presented aerospace concept is ultimately doomed, whereas a technically perfect but badly presented concept is immediately doomed. I don't know how much the picture helped, but it certainly didn't do any harm.

..............................................................................................

There was only one way to find out for certain whether *Huygens*'s entry could be seen from Earth, and that was to conduct the experiment. The good news was that those allocating time on the HST were persuaded it was a good idea. The observations would provide some useful additional data about Titan too.

The bad news was that the HST observation would, in practice, never take place because of a technical failure with the instrument that was to have been used.

..............................................................................................

### RALPH'S LOG, 2004

#### HST CYCLE 13

*January.* When it came time to actually propose the details of an observation, a year before the encounter, I teamed with Keith Noll, a veteran planetary astronomer who, based at the Space Telescope Science Institute, would know how to get our rather awkward and special observation implemented. Mark Lemmon, with whom I'd worked on most of the Hubble observations of Titan over the years, would be a vital ally. And Jean-Pierre belonged on the team, acting as conduit for the final tracking information to target the observation. Plus, with Hubble being a joint NASA-ESA project, having a European coinvestigator wouldn't hurt politically.

The proposal was pretty slick, I thought. We'd ask for one and only one HST orbit, since these things are like gold dust. Plus, having served on the solar system science panel for HST, I knew that the end of the allocation process was a bit of a knapsack problem—after the front-running handful of proposals had been selected, there might be one or two orbits left over, with the next-highest-ranking proposals needing maybe four or six. So a proposal needing only one orbit might sneak in, just because of the statistics of small numbers. Our proposal might fit into what was left, but a higher-ranking but bigger proposal wouldn't. Risking one orbit on the uncertain chance of detecting the entry was probably all that prudence would allow anyway.

We designed our observation to give good Titan science anyway, even if the entry wasn't detected. We'd be able to measure the north-to-south variation in Titan's haze and detect the polar hood (if it was still there), and the spectrum of Titan at the probe descent region we would obtain from above would be a useful dataset to compare with the probe's observations from below.

Happily, our proposal was selected. Of course, this just means more work—the detailed timeline of the HST orbit would have to be choreographed, which meant installing and learning HST's sequencing software. This constructed a timeline that minute-by-minute accounted for all the steps of zeroing in on Titan, taking the sequence of spectra and allowing for the precious seconds spent reading the data out of the instrument.

*Summer: Bad News from Orbit.* The rumors reached me before the official e-mail did. In August 2004, a five-volt power supply in the space telescope imaging spectrograph (STIS) had failed in orbit. "Switch to backup" is the usual response of a starship commander to such a problem. However, this already was the backup—the prime five-volt supply had failed long before. So STIS was dead.

And every observation proposing to use STIS was dead too—including mine. The first and only successful HST proposal I'd ever written (and judging from the steady decline of the telescope and the hiatus in shuttle flights after the *Columbia* accident, quite probably the last), and it was torpedoed by some little circuit burning out.

Having been on an HST panel, I knew how many excellent proposals using other instruments had been rejected, and so there was no danger of HST itself being unused. There was to be a process of reevaluating observations. If a program's goals could be met with another instrument like the advanced camera for surveys, then it would use that. But otherwise, the observation was lost and the time would be allocated to another proposal.

We looked into it, but naturally ours was impossible with another instrument. We relied on the spectrometer to pick out the faint filigree spectral signature of CN violet emission. A camera with a broad filter just wouldn't do it. All that effort wasted, and perhaps the best chance of detecting the entry gone. That's show business.

..............................................................................................................

With the probe released to its own fate, the *Huygens* science teams had a week or so of respite before attention would shift from sunny California to chilly Darmstadt, Germany, home of ESA's European Space Operations Centre (ESOC).

# 5. Landing on Titan

Darmstadt, a short distance south of Frankfurt and its massive airport, is an unobtrusive little German town, perhaps an unlikely home to the European Space Operations Center (ESOC). But this is where the data from *Huygens* were to be received, and in the week or so before the probe's descent onto Titan on January 14, 2005, the science teams began to gather in anticipation.

......................................................................................

## RALPH'S LOG, JANUARY 8, 2005

### Tucson, Entry Minus Six Days

I am in my cluttered office in Tucson. I like to think with all the mountain of papers and books here that I am doing my bit for carbon sequestration, fending off global warming by at least a little. There has been enough snow in the last few days to permit skiing on Mt. Lemmon, something I would normally take a day off to do, but somehow the risk of breaking a leg just before the culmination of a career's work makes me chicken out.

I am packing for Germany and the probe descent. I try to guess what I'll need. The real analysis of the data will take place over the months and years after it is received, but in the hours and days immediately afterward, there will be the need for quick and dirty answers, and we'll need background information. Which temperature sensor was closest to the SSP electronics box? What is the density of liquid ethane? E-mails from my colleagues on the SSP team assure me that all the SSP calibration info will be to hand. I try to think of what is not obvious to bring, what we could possibly need that everyone else doesn't think we'll need. Half reluctantly I pack the bulky *Huygens* user manual—someone in ESOC somewhere will have one, but perhaps not in our room, not to hand. Most of the other information is, I am sure, on my notebook computer.

I also fish out my test penetrometer—built at a different time with slightly different cables, but made to the same pattern, and machined out of the same piece of titanium with a piezoelectric disk from the same batch as its cousin, now a billion miles away and coasting quietly inward to Titan. Assuming airport security doesn't confiscate it, it might make a useful prop for media demonstrations if we get data back from a solid surface.

I realize that I don't recall where, years ago, I put the diskette of my *Huygens* crash simulator, to help translate the impact deceleration signature into mechanical properties of the soil. It is software I basically haven't touched since I wrote it in 1993—in GWBASIC on a DOS PC. Perhaps if we get data on this event, I'll rewrite it better, and in a more modern programming language. But for now, the old program will have to do. I made sure my current notebook has BASIC on it (not actually a trivial issue) so the code will run, but where is the code?

Rather than turn my office upside down looking for a diskette that I may or may not find and may or may not be readable anyway, I grope under my desk and pull out my old laptop from which I knew the code hadn't been deleted. This beast—back when laptops really were laptops needing a solid pair of femurs beneath to hold them up, not the handy "notebook" of today—was the machine I wrote the code on as a PhD student. Perhaps in anticipation of this very moment, I had never got around to disposing of it. The battery had long since failed, but it turned on first time, its big British power plug hanging precariously on an adaptor to the U.S. AC outlet. The internal lithium battery that keeps a memory alive has also died years ago, and so the boot-up process is interrupted—the machine needs to be told what kind of hard disk and floppy it has (this is before Windows and Plug and Play). Prompted by a Post-It note above the keyboard, I enter type 42 for the type of hard drive—a whopping 40 MB—and then enter a 4 for a 1.44-MB floppy. Ye gods, was there ever another kind of floppy disk? I vaguely remember something about double density versus high density. Anyway, reminded what its own faculties are, the machine fires up with a beep.

I navigate quickly to my old directory and copy the various versions of the program, imaginatively titled "CRASH1.BAS," "CRASH2.BAS," and so on, to a floppy, which groans and grinds but finishes OK.

On my office PC, I copy the files onto a dinky little memory stick the size of my thumb, costing thirty dollars yet possessing the storage capacity of a whole crate of floppies. And I'm ready to leave.

I suddenly realize that, depending on what happens next week, the reasons for hanging on to a lot of these papers, and the old laptop in particular, will be gone. Maybe we'll land on a solid, and my decade-plus interest in splashdown dynamics will have been for naught. Maybe we'll get no data at all. One way or another, a major spring cleaning will be in order. Many of my own memories, too, have been formed in this project. Perhaps, when I return in ten days, I will be a different person—reformatted by the anguish of a lost mission.

The old hard disk on the laptop whirrs noisily. Even though I doubt I'll need it ever again, it seems disrespectful to switch the machine off without parking the disk, which had to be done "manually" on machines this old. Another Post-It reminds me what to do. I type "BYE."

..................................................................................................

As the final few days ticked away, ESOC became busier and busier. More and more scientists appeared. Some of the first to arrive were the junior but no-longer-so-young "regulars" who, for all the years of waiting, had reported their instruments' status in orbit or laboratory analyses, or had helped develop the workarounds to the various problems that had cropped up. These foot soldiers, approaching their finest hour, prepared the ground for the looming battle—getting computers set up, getting the printer working, learning the planned sequence of events. Toward the end of the week, many others elbowed their way into the PISA (the Principal Investigator Support Area, a big open-plan office made temporarily available)—some with key roles defined, some being scientists who had been involved in the original proposals or early development, and some VIPs associated with experiments in some political context, keen not to miss the space spectacle of the year. A few journalists also began to trickle in.

ESA arranged a dinner for the project personnel and their guests. It was a welcome reunion for current team members and former colleagues, many of whom had moved on to other projects or had retired.

## THE PLAN FOR *HUYGENS*

At this stage, there was nothing the mission teams could do on the ground to influence the events that would shortly unfold on distant Titan. The hope was that all the commands loaded on *Huygens* and *Cassini* would be executed according to a plan largely shaped years earlier.

*Huygens* would spring into life when it was a few tens of thousands of kilometers above Titan and traveling at a speed of about 6 km/s. The computers would boot up and instruments would be turned on. The thin atmosphere would start to have a measurable effect on the probe at an altitude of 1,000 km or more, even though the atmosphere is a billion times thinner there than at the surface. *Huygens* would begin to be slowed down by the drag of this thin air on its blunt, conical heat shield. Scientific measurements would commence. The shiny foil coating that protected the probe from the Sun's heat when *Cassini* was at Venus would be torn from the outside of the heat shield as *Huygens* plunged deeper into the atmosphere. Peak heating would be at an altitude of about 400 km above the surface. At this stage, *Huygens* would have barely slowed down, and the air in the shock wave in front of the probe would reach a temperature of 1,400 K. Even the back of the probe was covered with insulation to protect it from the glow of the hot air.

*Huygens* would feel its maximum deceleration at a height of around 250 km, a few seconds after peak heating had occurred. Although moving more slowly than it had been, the probe would be in denser air and feel a deceleration of about 15 g. This was, however, quite modest by planetary probe standards. About a minute after this peak of deceleration, the probe would be down to an altitude of about 170 km and its speed reduced to about 350 m/s, which is about one and one-half times the local speed of sound. The effects of breaking the "sound barrier" apply whether speed is increasing or decreasing; the fairly flat *Huygens* probe would have begun tumbling as it passed Mach 1, so the deployment of the parachutes would begin here. A mortar on the back of the probe would shoot a small "drogue" parachute through the probe's back cover. This parachute would both hold *Huygens* stable and pull off its back cover. As it came off, the back cover would pull out the large main parachute, some 8 m across.

With the deployment of the main chute, the front heat shield would fall away and the probe would decelerate quickly to a steady descent rate of about 50 m/s. A cover that had protected the DISR (descent imager/spectral radiometer) from material coming off the heat shield would fly off as its springs were released. Caps on the inlet pipes of the GCMS (gas chromatograph/mass spectrometer) would be broken off by explosive actuators. Two small arms carrying the electrical field sensors for HASI (the Huygens atmospheric structure instrument) would swing out and lock into position. Small vanes mounted around the edge of the probe would keep it spinning slowly so that DISR's cameras could pan around. (The line to the parachute had a swivel to prevent it from twisting up.)

The large size of the main parachute had been dictated by the need to pull the probe away from the heat shield safely. But if the probe were to continue to descend under the main parachute, it would take some five to eight hours to reach the ground, by which time the *Cassini* orbiter would have disappeared out of range before data could be taken from the surface. So, after ten minutes, explosive bolts would be fired to detach the main parachute. A smaller stabilizer parachute would take over and allow the probe to descend more quickly.

The ACP (aerosol collector/pyrolyzer) experiment would suck atmosphere in through a filter, trapping aerosol particles and cloud particles. It was thought that the tholin haze particles might act as the cores of small crystals or droplets of ethane or hydrogen cyanide lower down.

And at 40 km and below, methane condensation would be possible. Between there and the melting level at 14 km, a thin film of methane frost could form on the probe, although not enough to disturb its aerodynamics; all of these strange environmental effects had to be anticipated and their effects evaluated. A small oven would break down the material trapped in the ACP so that its composition could be analyzed by the GCMS.

As *Huygens* neared the surface, gently swinging under its parachute, it would be descending at about 5 m/s. Its height would be determined by a radar altimeter on board, and the instruments would adapt their operations to maximize the scientific return. At an estimated two minutes from impact, the SSP (surface science package) experiment would prime itself. Its acoustic sounder—like a small sonar—would begin sending out a rapid series of pings. A lamp would illuminate the surface in the last few tens of meters of descent, allowing DISR to take a surface spectrum despite the anticipated gloom below the haze.

*Huygens* had to be prepared for anything to happen at impact. Before the mission, no one had any idea what the surface would be like. It had not been feasible to design the probe so it would definitely survive, but a wide range of possibilities had been taken into account, including a splashdown in liquid, and there was reckoned to be a good chance that *Huygens* would keep transmitting data from the surface. After listening out for the probe until it was no longer visible, *Cassini* would slew around to relay the probe data back to Earth.

If nothing broke on impact, the probe would continue to take data, but it had not been designed to operate in any specific way after it had landed. There could be pictures, but they would be limited to wherever the camera happened to be pointing when it came to rest—even if that was the underside of the parachute! There was a chance that some surface material would be vaporized by the heated inlet of the GCMS so that it could be analyzed. After a landing in a liquid, the acoustic sounder on the probe could possibly have picked up an echo from the bottom and hence determine the depth.

In the end, the collection of data would cease for one reason or another. The batteries would run out, or something would stop working as the probe cooled down. Even if *Huygens* continued to transmit, *Cassini* would eventually move out of range of the probe. The data link would fade and then disappear. Several hours after the end of the probe's mission, *Cassini*

Figure 5.01.  The *Huygens* encounter was a major media event. Project scientist Jean-Pierre Lebreton announces the status of the probe to the assembled hordes of media. In the background is a full-scale model of the *Huygens* probe (SM2) that had been used in parachute tests, dropped from a helium balloon 40 km above Sweden in 1995. (ESA)

would direct its antenna toward Earth and download the data it had collected from *Huygens*.

All this, at least, was the plan as articulated in everything from detailed specification documents to computer animations put together by ESA and others for media use. But nothing ever quite goes to plan.

### LANDING DAY

By the time the day of reckoning arrived, the media presence had become almost oppressive from the point of view of the scientists. The European Space Agency was pleased with the level of interest, although its public relations staff numbered only a handful, and the intensity of the event threatened to be overwhelming. About eighty busy scientists were being besieged by some two hundred journalists, reporters, and camera operators.

As one might expect in Germany, the logistics were well arranged. While TV trucks clogged the roads on-site, parking passes were provided for the scientists. The activity would be centered on two rooms at ESOC, the PISA and the MCR (Main Control Room)—the usual sort of control room with lots of screens, where the first news would come in and so all the VIPs could appear to be involved.

The PISA was a big open-plan office: indeed, an agenda item of many *Huygens* meetings in the previous year was on encounter logistics. How many tables would be needed, how many Internet connections? All these details, mundane as they are, need to be worked out in advance. Access to the various areas was controlled by swipe cards, and with few occasional exceptions, the media were, thankfully, excluded from the PISA, so the real work would not be hindered. The number of people involved meant, sadly, that not everyone would fit in the PISA, and the large DISR team was therefore secluded in a Portakabin a few minutes away.

...............................................................................................

### RALPH'S LOG, A.M., JANUARY 14, 2005

It was going to be a long day—most people arrived around 9:00 or 9:30. Although the probe data would not come in for hours, the phone rang continually with requests for media interviews. I checked up on the ground-based observing campaign: snow and high winds on Mauna Kea in Hawaii were making life difficult for the astronomers. Someone checked his watch and observed that even though the light from the event would take another hour to reach the Earth, *Huygens* had already encountered Titan.

The HST observation was, of course, not happening. But the Keck and Infrared Telescope Facility (IRTF) telescopes in Hawaii—on the other side of the world— were acquiring images and spectra, respectively, of Titan as the probe entered and descended. The Gemini telescope could not open its dome in the high wind. Meanwhile, in California, the Palomar Telescope (the only large telescope that was set up to observe in the near-UV) was unfortunately clouded out.

...............................................................................................

Around 11:40, a cheer went up. There was news. The Green Bank Radio Telescope in Virginia had picked up a signal from *Huygens*. Whatever happened next, the mission had not been lost without a trace, like *Beagle 2* a year before. This signal meant that the back cover was off and a transmitter was on. *Huygens* had survived its fiery entry into the atmosphere. Presumably, the parachute was out and the probe instruments were working. Detecting the signal on the ground had always seemed like a long shot, but there it was, clear as a bell.

Figure 5.02.  *Left*, the Green Bank Radio Telescope in Virginia, which eavesdropped on the *Huygens* transmissions to *Cassini*, and thereby provided the first news that the probe was operating successfully. *Right*, a dynamic spectrum or "waterfall" chart, like those used by sonar operators aboard submarines. Brightness represents intensity against time (vertical axis) and frequency (horizontal). Random noise appears as spurious dots, but the vertical line denotes a steady signal—that from *Huygens*. (NRAO)

Excitement started to build in earnest in the afternoon as *Cassini* began relaying the *Huygens* data to Earth. But something wasn't right, and as the *Huygens* data began to be read back, word began to circulate from the Mission Control Room about a problem with "Channel A." *Huygens* had two separate computers, each wired to a separate radio transmitter, A and B. *Cassini* carried two separate receivers, so that a single failure anywhere in the system would not cause the mission to be lost. The transmitter and receiver on Channel A were equipped with both ordinary oscillators to control the radio frequency and special ultra-stable oscillators (USOs). The purpose of the USOs was to make Channel A exceptionally precise so that the probe's motion could be tracked by using the Doppler effect. It turned out that the problem was with the *Cassini* orbiter. Although everything on the probe was working, on the orbiter the receiver had correctly been commanded to *use* the USO, but the command to *switch on* the USO had been missing. It was a simple, but catastrophic, mistake. Without the USO on, the receiver could not find the probe's signal.

The Doppler Wind Experiment team was devastated by the loss of its data. Others wondered what else might have gone wrong, and what the loss of Channel A would mean for their investigations. Some experiments duplicated the data on both channels, to be more certain that all the data would be received. Others took a gamble and made the most of the two channels by sending different data on each. If both worked, they'd get twice the amount of data. (Each channel supported eight kilobits per second, about seven times slower than a turn-of-the-millennium-era dial-up modem. The total return over three hours would fit easily on a CD-ROM.)

Late in the afternoon, the readout of the probe data from the Jet Propulsion Laboratory (JPL) to ESOC and the distribution of the data began. The stream of information, recorded chronologically by *Cassini*, had to be chopped up and the right packets of information sent to the right teams. This process took an hour or two, and while the experiment data accumulated, only a few pieces of engineering data could be monitored. But these seemed to offer good news—not least that over an hour of data was received after the probe had landed. The landing was quite late: two hours twenty-eight minutes after the parachute was deployed. The nominal descent was supposed to be two hours fifteen minutes, and no more than two hours thirty minutes.

But it was already known, from the continued detection of a radio signal from *Huygens* by the Parkes radio telescope in Australia (after the Earth had rotated such that Titan had set below Green Bank's horizon), that the probe had continued to transmit for several hours after impact!

........................................................................................................

### RALPH'S LOG, P.M., JANUARY 14, 2005

It is a tense wait for the data to arrive. Happily, there is an excellent coffee machine installed in the PISA. I fire up the game Mechwarrior on my laptop and destroy some robots; there didn't seem to be anything else better to do in between answering calls from reporters. There wasn't much to say to them—it seemed the probe had worked. The fact that it had apparently kept operating for so long on the surface suggested it had probably landed on something dry. To have anything else to say, we'd have to wait for the actual data to be distributed.

Around 6:00 p.m., I get called away for an interview with ABC Australia. When I return to the PISA, the SSP team members are intently hunched over laptops. Gaaaah! The data have arrived, and I missed the jubilant moment!

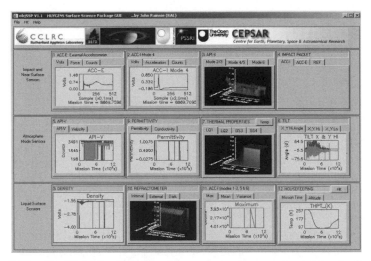

Figure 5.03.   The "Quick Look" display of the surface science package data—the first view we had of what happened to the probe (impact plots are at upper left). Much of the other data took weeks or months to fully interpret.

"Who's got the stick? Gimme the f—ing data," I blurt, desperate curiosity getting the better of politeness. After an agonized minute, I am presented with data that this morning was transmitted from the probe on Titan, a billion miles away.

After descending a rather convoluted directory structure, forced by a last-minute adjustment to the display software, I open the file o518/o518/o518ss/o518_IDL/o518ss_chB. (Stream 518 was the telemetry package from the mission. Later streams would patch up small errors. There had been many previous streams from the in-flight checkouts.)

The penetrometer record looks weird. There's a suspicious big spike at the beginning. Maybe it broke. In fact, everything looks weird. The speed of sound profile doesn't look right. The tilt sensor record is all over the place, looking like a seismogram instead of the record of gentle swinging that we had expected.

But the world wants answers, so we have to focus on our assigned tasks. Andrew Ball and I have the job of making sense of the impact measurements, ACC-E and ACC-I. Having the two was a hedge against the unknown. If we splashed down into a liquid or landed on something very soft, ACC-I would catch the event but ACC-E would be useless; if we landed on something hard, ACC-E would come into its own, but all ACC-I would tell us would be how hard the probe resisted as it crumpled on impact. For some range of surface properties in between, both sensors would tell us something— and in fact, that was what had happened.

I look at the ACC-I record. At least it looks like a real impact profile: 15 g—that's not much, but too hard to be a splashdown. A pity—there goes my bet. Some high school mathematics suggests that the probe came to a halt over a distance of about 15 cm. Knowing the shape of the bottom of the probe allows an estimate of the contact area, and thus the mechanical strength of the surface material. It was soft, like packed snow or sand, or wet clay. I compare the shape of the deceleration curve with simulations I had done with my computer model all those years ago. It doesn't quite look like the dry sand models. Dry sand, unless it has been packed down, is a little fluffy and so needs to be compressed some-what before it begins to resist penetration, and the deceleration record seemed to jump up too quickly for that. So a wet clay or packed granular solid seems to make most sense. The HASI team is on the other side of the table. We check with them that the peak impact deceleration they see is about the same as ours—a reassuring Italian shrug confirms 15 g.

We then devote our attention to ACC-E. The big spike at the start was a distraction, and a broad mound on the profile seems to coincide with when the base of the probe would have hit the ground. (ACC-E had been tested numerous times in the laboratory, but not with a $300 million, 200 kg space probe attached.) Dismissing those two anomalies, we could see that the force profile was pretty flat, about 50 N. So the penetrometer had sensed a force much as if you had pushed your finger into a material with about one-twentieth of your weight, say 5 kg, behind it: packed snow, wet clay, maybe rather stiff molasses.

But what is the spike? Had something broken? We reason that perhaps there could have been some anomaly (like an electrical spark between the ground and probe at impact due

Figure 5.04.　The surface science package team's corner of the PISA, with the first author in center. The level of crowding is evident. This was not an ideal environment for quiet contemplation of the new data.

to charging by cloud droplets, just like early aviators used to get with rubber-tired planes before it was realized that a conductive skid was better). But otherwise, there was something with more strength or inertia at the beginning of the impact record—a hard crust, or a pebble.

I am afraid I don't remember at what point the first DISR pictures appeared, showing the cobbles littering the landing site, looking like a streambed or a beach with the tide out. But at least the first analyses of the impact data were made "in the dark," with just a squiggly line on our screens, no hint of what the images would show.

We have to prepare two charts of results within the first couple of hours. The project is hungry for results, the media hungry for news. We list the possibilities, and playfully suggest "crème brûlée" as an analogy for the surface. (It may have been Andrew who first spoke the words, one of my favorite desserts, aloud. Even though the findings are under strict em-

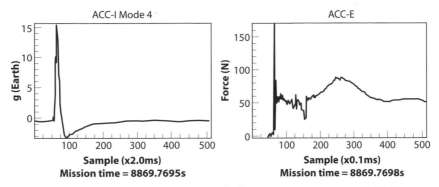

Figure 5.05. The impact deceleration record (*left*) and penetrometer signal (*right*). The 15 g impact lasted less than one-tenth of a second and suggested a somewhat soft surface, whereas the penetrometer suggested a lump or a crust above a soft surface—hence, the "crème brûlée" analogy. The penetrometer gave the most accurate determination of the landing time.

bargo, I send an e-mail containing only those two words to Zibi back in Tucson—she'd know what it meant!) We debate whether we should keep it on the chart, but it fits the data, and it seems to conjure up an exotic appeal, so we leave it in. We then have to spend about ten minutes Googling on the Internet to make sure we have got the French accents right.

When the SSP principal investigator, John Zarnecki, spoke at the press briefing, the impact results formed the centerpiece of the SSP quick results, since the other data would take longer to interpret. He dutifully read the possibilities, and the media latched onto the food analogy. "Titan team claims just deserts as probe hits moon of crème brûlée" read the headline in *Nature*.

Though every team was busy making sense of its own data, the highlight results were going to be the DISR pictures. Each individual image was very small by modern standards, about one-twentieth of a megapixel, and highly compressed at that. But they were striking nonetheless. The first images that came up on the DISR team's screens were the most surprising—the knee-high view of Titan's surface showing a scene littered with rounded "rocks" on a smooth plain.

This was the first image that was released to the public—sometime after 9:00 p.m. local time. As Emily Lakdawalla of the Planetary Society (who was "embedded" as a reporter at ESOC to cover the event, as well as to present the results of an art competition to portray how Titan might look) observed, "Any geologist worth her salt thinks of one thing and one thing only when she sees round rocks: some river of some liquid has rolled broken chunks around, wearing down their edges, making rounded cobbles." This was not something anyone had dared predict!

In the second of the three images shown (this one from some height above the surface looking down), the picture of Titan as a river world was reinforced. A bright highland area was dissected by a network of what looked like dark river channels, branching and winding just like rivers on Earth. And they seemed to drain toward a dark, bland region, with the boundary between them fairly sharp, like a coastline. The cosmopolitan group of scientists saw different analogues—some saw the French Riviera, some saw the California coast. But wherever it most resembled, it looked very Earth-like.

A third image was a little inscrutable. This one, from higher altitude still, was much harder to interpret. Not only was the contrast lower so that noise in the image (from the data compression process, for example) made it harder to see detail, but the features themselves were much less familiar. There were many arrow-shaped bright features, some connected together. Even a couple of years later, it isn't quite clear what these are.

························································································

### RALPH'S LOG, JANUARY 14–15, 2005

Remarkably, within an hour or so, the efforts of various amateurs around the world, pouncing on the DISR images released on the Web (the dataset was released rather quicker than had been the original plan), started to appear. Some impressive mosaicking efforts were made in the following hours and days; notably, it was only an hour or two before some bright spark had Photoshopped the "Face on Mars" onto one of the Titan images. It got pinned up near the PISA coffee machine, which was running in overdrive, a mountain of used coffee cups building up next to it.

Figure 5.06.    Among the first images to be released on January 14–15, 2005, was this small mosaic of downward-looking images (*left*) showing a bright highland dissected by river channels next to a dark, flat lowland. Perhaps the most remarkable image is at right, the unexpected view from the surface, showing a plain studded with rounded cobbles. (ESA/NASA/University of Arizona)

It is near midnight U.K. time—but apparently some half a million people are watching live coverage on TV. On a Friday night, I have to wonder how many of them are sober—but what the hell? I get my fifteen minutes (actually about forty-five seconds) of fame, explaining what the penetrometer did. Wow. It dawns on me that eleven years after I spent three years specifying, designing, and building it, a billion miles from Earth and 180 degrees below zero, the thing had actually worked for the one-twentieth of a second it was supposed to!

The SSP team had a sweepstake running on the descent time (defined as impact determined by ACC-E relative to T0, the moment at which the parachute mortar had been fired). My guess had been two hours twenty minutes, slightly longer than the nominal expectation. I learn that my chance of winning had, in fact, been tiny—my guess had been closely bracketed by others. It is principal investigator John Zarnecki, who had been my PhD supervisor all those years ago, who guesses impressively close to the real descent duration: two hours twenty-seven minutes (specifically, 8869.7598 s). He denies any inside knowledge; good instincts are just part of the package of skills for a successful PI.

Zarnecki graciously opens his sweepstake prize, a bottle of sixteen-year-old Lagavulin malt whisky. Those of the SSP team still around, plus a few of the TV crew, savor it. What a day!

I get back to the hotel at 2:00 a.m. but am too buzzed to sleep. I work on making a simulated sonar sound from the echo records, which in the end does not get used. But who cares? It worked! The probe had worked! We had got to Titan.

....................................................................................................

## THE DAY AFTER

The probe encounter itself was attended by ESA's director of the scientific program, David Southwood, and by his NASA counterpart, Al Diaz. Both appeared to find it an emotional experience during the massive press conference on Saturday, January 15, the day after arrival. Diaz's tearful praise of the teams' efforts that "culminated in this one moment in history—it's just incredible" earned him a mocking on Comedy Central's *Today Show*. Southwood, who had been the original team leader for *Cassini*'s magnetometer experiment until he was recruited in May of 2001 to be the ESA's science director, wistfully recalled a poem of exploration by Keats.

Southwood acknowledged that there had been a problem with the radio transmission from *Huygens* and that there would be an inquiry. But no one really cared that much; the mission had basically been a success. The hundreds in the audience, journalists and scientists, just wanted to hear some new results—even though it had been only about fourteen hours since the raw data had hit the ground.

Marcello Fulchignoni offered some ear-catching (although scientifically rather meaningless) sounds—one the noise of the wind rushing past the probe as it descended, and the other a modulated techno buzz generated from the radar altimeter signal. The crowd loved it. He also reported the surface conditions at Titan: 93.6 K and 1.46 bar.

Marty Tomasko presented a preliminary mosaic of the images. Although there was no evidence of present-day liquid exposed on the surface, it seemed clear that flowing liquids—and apparently liquids dropped from the sky—had carved channels on Titan's surface, and rolled materials around. The results of the GCMS instrument also seemed to suggest that methane was abundant at Titan's surface.

After the adrenalin-filled late-night rush of the encounter itself, and the press conference (and many interviews) the following day, everyone was burned out. It was time to start a more considered analysis of the data.

On Monday or Tuesday, three or four days after *Huygens*'s successful descent onto Titan the previous Friday, the fuss surrounding the event itself was dying down. The media circus fizzled away, and VIPs found other places they needed to be. ESA was to hold a press conference at its Paris headquarters on Friday, so there was some late juggling of itineraries on the part of the senior figures summoned to appear there. But most team members were heading back to their usual places of work to begin the serious task of interpreting in detail the wealth of data the experiments on *Huygens* had returned from Titan.

....................................................................................................

### RALPH'S LOG, JANUARY 28, 2005

#### Pilgrimage

A couple of weeks after the encounter, I visited my SSP colleagues at the Open University in England. En route, though, I made a stop at the London Science Museum. Among the many neat exhibits and demonstrations are formidable steam engines of the Industrial Revolution, a replica of Babbage's Difference Engine, and a selection of aircraft and spacecraft, with the beautiful but petite Black Arrow rocket hanging from the ceiling. For a month or two, there is a special exhibition relating to *Huygens*—I have to go and see it. And there, in a glass case among all these marvels of technology, is something I made. It is the flight model ACC-E penetrometer (actually mislabeled as an engineering model). This is the unit I had assembled over a decade before, the one that fit together the most perfectly, the one that was supposed to go to Titan. But months after I had put it together, a technician had tightened it the wrong way, cracking the ceramic force transducer.

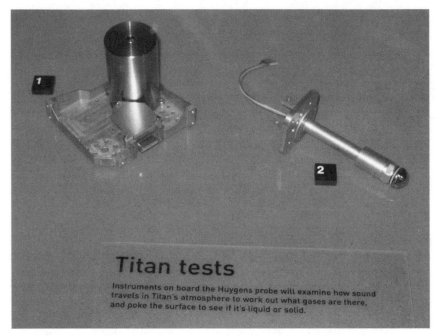

**Titan tests**

Instruments on board the Huygens probe will examine how sound travels in Titan's atmosphere to work out what gases are there, and poke the surface to see if it's liquid or solid.

Figure 5.07.   SSP hardware in the London Science Museum. With the mission now history, hardware is consigned to the museums. These are spare parts of the thermal properties sensor (*left*) and the penetrometer (*right*). The penetrometer head is 14 mm in diameter. The unit on display was actually built by the first author as the flight model to go to Titan, but had to be replaced with a spare before launch.

And so, what had flown to Titan was the backup we had made, the flight spare. It is a good lesson for soccer substitutes or acting understudies: one day the call may come to step up into the limelight. Second best as it had been, the flight spare had done well.

..................................................................................................

## A CASE OF SPIN

After a couple of days, some disturbing and confusing aspects of the descent began to emerge. The tilt sensor data from the SSP looked rather violent, though it would become clear later that the tilts had not been nearly as bad as the readings suggested; it was more a case of sideways buffeting than rocking as such. One of the tilt sensors was arranged radially on the probe, such that the probe's spin should fling the little slug of liquid outward like water in a washing machine. The tilt readings were the position of this slug, as measured by a pair of electrodes. So in principle, if the probe were doing nothing but descending steady and flat with a uniform spin, the tilt reading would be fairly constant, corresponding directly to the spin rate. Amid all the oscillations and noise, it was hard to see, but by averaging dozens of readings, it seemed at least more consistent with other data.

But at the beginning of the descent, the average tilt value was puzzling. It was in the opposite direction from where it sat for most of the descent. And the tilt was negative—but that made no sense. It wouldn't matter which way the probe was spinning; the tilt should be positive. The only way to get a negative tilt from spin was if the spin axis were outside the probe. Perhaps the parachute was doing something unexpected, swinging the probe around in a conical motion.

At that point, the DISR team had an even stranger suggestion. As they began to piece together their jigsaw of images, it became apparent to them that, for most of the descent, the probe appeared to be spinning backward, not in the direction it should have been! This suggestion was not greeted with enthusiasm by the industrial team; such an anomaly would need some explanation.

Insight into what was actually going on came sometime later from a piece of housekeeping data that turned out to be far more rewarding than anyone expected: the automatic gain control or AGC loop on the Channel B *Huygens* receiver on *Cassini* (Channel A, of course, being out of action). The AGC was a circuit that compensated for the varying signal strength to maintain a more or less constant output in the radio receiver. Thus, by monitoring the state of the AGC, one could get an indirect measurement of the signal strength. This was recorded about eight times per second, and showed a rich variation during the descent. The spatial pattern of radio emission from the probe antennas, which were on the roof of the probe among the parachute boxes and other items that cause all kinds of reflections, was not uniform, but looked like the petals of a flower, with stronger signals in some directions and weaker ones in others. As the probe spun, these petals swept around, and when a strong lobe in the pattern pointed at the *Cassini* receiver, the signal was strong. The signal strength went up and down like a heartbeat in a repeating but slowly evolving pattern.

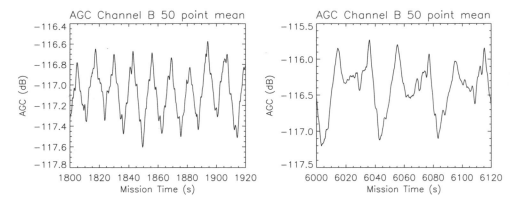

Figure 5.08. The "heartbeat" of varying signal strength received by *Cassini* as the probe rotated under its parachute. Both plots are about two minutes long. During the first (*see left*), half an hour after parachute deployment, the probe is spinning about 5 rpm, whereas an hour and one-half later (*see right*), it has slowed to about 2 rpm. The slight differences from one cycle to the next are partly due to swinging and buffeting of the probe.

A week after encounter, a sound file was made out of these data. It was a hypnotic sound, with beats that faded in and out as different swinging modes took over and the spin rate evolved. A graph is not always an appealing mode of presentation, and sounds have their uses, not least for radio news.

Ultimately, analysis of the signal strength changes would show conclusively that the probe had been spinning the "wrong way." After beginning its descent under the right conditions, it had apparently spun down to nothing and then spun up in the opposite sense. It seemed incredible. Even as this book goes to press, exactly why the probe spun anomalously despite the presence of the spin vanes is not understood. Whatever the explanation, it wasn't simple. Some detective work later confirmed that the spin vanes were put on the right way!

## DISR: *HUYGENS*'S IMAGES AND SPECTRA

Another surprise was that the high-altitude images were so indistinct. All the models of Titan's haze had indicated that the haze should "clear out" below an altitude of perhaps 80 km. Titan's reflectance spectrum does not fit with a model of how light at different wavelengths is scattered and absorbed (most notably in the 619-nm methane band) unless haze is absent at low altitudes. But this belief, held pretty universally, relied on some assumptions about the scattering properties of an individual haze particle that turned out to be unjustified. And these assumptions, Marty Tomasko explained later, were due to computational limitations in performing the modeling.

Many haze models assume that the haze particles are small spheres ("monomers"), but *Voyager* measurements had shown that a single population of spheres did not fit the data. This led to more sophisticated models, which assumed the haze particles to be aggregates of smaller particles (perhaps spheres). These aggregate particles have some of the optical properties of the monomers and some of spheres close in size to the aggregate as a whole. But calculating the electromagnetic interactions of all these tiny spheres is a formidable computing problem, which increases significantly in complexity according to the number of spheres assumed to make a particle. For this reason, the maximum size of aggregate particles in the pre-*Cassini* models was restricted.

The *Huygens* data could, it seemed, be fit only with much larger aggregates, with hundreds of monomers. This increase didn't make the particles that much bigger: doubling the number of monomers, for example, causes the diameter of the aggregate to increase by only 20 to 40 percent. The change, however, made it possible to accommodate the ground-based observations and yet still have haze extending down to the ground, rather than being washed out in the lower atmosphere.

The DISR team had devoted many person-months of effort to preparing software to be able to process their data immediately when it arrived—not just mosaicking the images together, but analyzing the way sunlight was scattered by the haze. However, while mosaicking could be done by hand (as indeed many amateurs showed by making impressive mosaic products at home), the quantitative analysis of spectra to determine the haze properties required knowledge of the probe's orientation at the instant the data were acquired.

Most important in this knowledge was the azimuth angle—the rotation of the probe about its central axis (which, by and large, was close to vertical throughout the descent). In theory, this was to be measured with ease and precision by a Sun sensor in the DISR instrument, which generated a pulse when the shadow of the Sun cast by a bar above the sensor head passed across a photodiode. By measuring the interval between these shadow pulses, the DISR instrument could measure the spin rate and interpolate the spin phase, thus giving information on the azimuth.

But several factors conspired to thwart this elegant plan, which made most efficient use of the telemetry bandwidth available by taking images and spectra only at the desired azimuths. First, rapid oscillations of the probe, and what turned out to be its reversed spin direction, misled the onboard algorithm. Also, although the signal from the photodiode was adequate at the beginning of the descent, a subtle combination of the drop in temperature of the sensor, the narrow range of wavelengths it responded to, and the variation of sensitivity with temperature conspired to suppress its signal at lower altitudes. This is the sort of effect that only the most exquisite and expensive testing on the ground would have discovered. But many of these details did not emerge until after some weeks or months of analysis.

However, the random timing of images did give rise to some neat results. Some terrain was imaged several times, from different altitudes. By comparing these images as stereo pairs, Larry Soderblom was able to construct a digital elevation model, showing in detail the surface topography. One such area was the bright highland with its river channels. The gullies were very steep, it turned out, and the bright area stood perhaps 100 m above the dark plain on which the probe landed.

## RECOVERING THE DOPPLER WIND EXPERIMENT

The total loss of the Doppler Wind data from *Huygens* because of the error with Channel A initially seemed like an enormous blow. An attempt to salvage a measurement on *Cassini* of the Doppler shift on Channel B did not succeed (nor was it expected to), because the drifts and jumps in frequency due to the much less stable quartz oscillator used on that channel made it impossible to attribute the measured changes in frequency to winds with any degree of confidence. But the objectives of the experiment were all achieved after all, thanks to remarkable observations made from Earth.

Almost all the ways of determining wind speeds involved measuring where the probe was at different times and inferring the winds from the probe's motion. But each of the methods was, in some sense, incomplete.

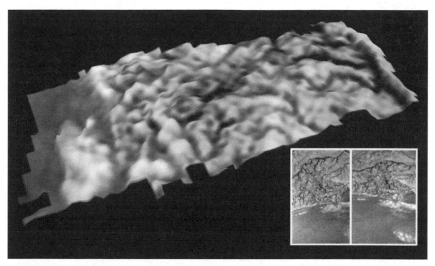

Figure 5.09. A 3-D view of a topography model generated by stereo comparison of the two DISR images in the inset. The bright area was inferred to be about 100 m above the dark plains. At the scale of the individual dark river channels, the terrain is very rough indeed. (NASA/ESA/University of Arizona/U.S. Geological Survey)

The original Doppler Wind Experiment (DWE) was to measure very precisely the Doppler shift of the radio signal on Channel A of the probe. The experiment relied on a rubidium oscillator—essentially an atomic clock—attached to the probe transmitter. There would be differences in the frequency with which the transmission was received due to various subtle factors, even including the curvature of space-time due to Titan's gravity. But the dominant factor was the "range-rate"—the component of the relative velocity between transmitter and receiver along the line between them. One factor taken into account when the entry point for *Huygens* was chosen was to make sure this line was well positioned in a more or less east—west direction, so that the winds, which were expected to be zonal, would produce a strong and easily measured range-rate.

The Doppler Wind team, heartened by the experience of radio astronomers in detecting the *Galileo* probe signal in 1995, had also lined up radio telescopes on Earth to pick up the *Huygens* probe signal. In fact, the *Huygens* signal would be much easier to detect than *Galileo*'s, even though the latter was only half as far away, because the *Huygens* transmitter sent a constant carrier tone as well as the data; this constant tone would be much easier to lock onto. The corresponding experiment without such

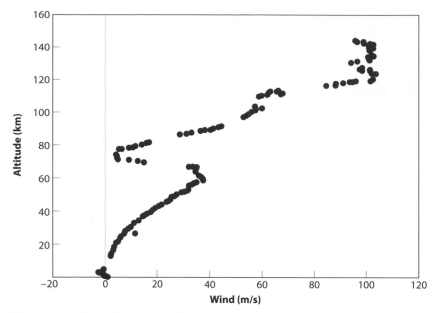

Figure 5.10. The profile of wind in Titan's atmosphere. The rapid high-altitude winds had been expected on the basis of *Voyager* and ground-based measurements. The rapid wind shear, leading to a layer of almost zero wind at 70 km, was something of a surprise, as was the small reversal in wind direction just near the surface.

a carrier on *Galileo* had required a vast postprocessing effort in order to pull the signal out from the background.

Thus, if all went well, there would be two independent measures of the probe Doppler shift. Because the line of sight to Earth was in a slightly different direction from the line of sight to *Cassini*, the combination of the two would allow the detection of any nonzonal (i.e., north–south) motion as well as the zonal winds.

In the event, of course, the *Cassini* measurement was lost. But the ground-based data were good enough to recover the bulk of what the onboard experiment was supposed to measure. The ground-based measurement was of lower resolution, the signal being far, far weaker, and there were a few gaps, but most of the wind profile was measured.

There were a couple of surprises. First, there was a layer about 80 km where the zonal winds—blowing at some 100 m/s higher up—dropped almost to zero before increasing again at lower altitudes. This strong wind shear may relate to the amount of sunlight absorbed at different altitudes. Second, in the lowest few kilometers of the atmosphere, the probe had drifted westward, not eastward as during most of the descent (as expected). This surprising reversal of the zonal winds at low altitude was confirmed by the DISR team, which could correlate the position of various surface features in images taken at different times—stereo in reverse—to derive the probe motion relative to them.

Another wind-drift experiment hadn't worked out. Thought up a couple of years before the encounter by Mike Allison of the Goddard Space Science Institute in New York, it was to have used the DISR Sun sensor to measure the change in the Sun's position as seen by the probe in order to determine how far the probe was drifting in the wind. It was an elegant idea, but the apparent failure of the Sun sensor to produce much data during descent meant it couldn't be applied.

Lastly, there was another experiment that was not really introduced until a couple of years before encounter. This was the VLBI tracking experiment. VLBI—Very Long Baseline Interferometry—is the technique of comparing the carefully referenced readings from widely spaced radio telescopes in order to determine the position on the sky of a radio source with great accuracy. The technique was used in 1985 to track the balloons dropped by the Russian spacecraft *Vega 1* and *Vega 2* into the atmosphere of Venus while en route to Comet Halley. The implementation for *Huygens*, though, would be much more precise.

As well as using special radio receivers attached to the radio telescopes and a formidable number-crunching effort afterward to correlate the signals, the technique also required some celestial points of reference. And so, over a year in advance of the mission itself, radio astronomers studied what was then to all appearances a barren patch of sky but would become the place where Titan would happen to appear, as seen from Earth, during the probe mission. In fact, seven months after the probe mission took place, some follow-up observations were made of one of the "*Huygens* Target Fields." These improved the accuracy with which a particular radio source (quasar J0744+2120) could be pinpointed relative to other sources in the International Celestial Reference Frame from twenty milliarcseconds to only one milliarcsecond. For comparison, Titan is about eight hundred milliarcseconds across as seen from Earth.

At Titan's distance from Earth, one milliarcsecond corresponds to about 6 km. The hope (the number crunching is still going on at the time of this writing) is that the position of the probe in the sky may be determined as a function of descent time to within a couple of kilometers.

This position is exactly complementary to the information derived by the DWE, and ultimately a combination of the two measurements should yield an accurate three-dimensional trajectory of the probe.

## THE RADAR ALTIMETER AND THE PHANTOM RAIN

Many of the experiments had little or no testing in exactly the way they were flown, the exceptions being the HASI experiment, which had wisely arranged several parachute drops from balloons over the years. And thus, the subtleties of the data took considerable time to think through, and phantoms of false conclusions flitted across the discussions before they could be reasoned away.

One of these involved the radar altimeters. They had worked relatively well, apart from a glitch that affected both. This caused them to make a false lock indicating half the real altitude for a while.

Roland Trautner of ESTEC on the HASI team, who had led much of the effort on the radar altimeters (and indeed had been in Brazil conducting a balloon and parachute test on them only six weeks before the actual encounter at Titan), noticed a curious pattern in the returns before they locked onto the surface. They seemed to be indicating a faint echo high above the ground. Cloud droplets would be too small to give a good echo, but raindrops would have the observed effect.

For some weeks the altimeter team thought they were on to something, but it was important to be sure because the probe was the only thing that had actually been to Titan, and whatever findings were reported from the probe would be taken as gospel for years to come. There were also some slight differences between the altitude reported by the altimeters and that inferred from the pressure sensors on *Huygens*. Roland went through his balloon test data and noted that the altimeters were affected slightly by changes in temperature. So then, some detective work was needed to work out the likely temperature history of the radar altimeter unit on *Huygens*. As luck would have it, the output from the temperature sensor closest to it had been on Channel A, and so was lost. Nearby sensors were compared with the thermal model and a "best guess" temperature profile was constructed. When that was done with some effort, the revised estimate of the noise levels in the instrument was able to explain what had looked like echoes from rain. It was a pity, but it underscored the need for careful contemplation of results from new instruments in unfamiliar environments before announcing what might be a major finding, or might just be a chimera like this. Ironically, a little over a year later, there were separate indications from other instruments that perhaps the probe *had* encountered some drizzle during its descent.

..................................................................................................................

## RALPH'S LOG, 2005

Just as NASA seems like a monolithic entity to outsiders when it is really a heterogeneous collection of fiefdoms, each with its own culture and style, people talk about "the media" as a whole, but the term hardly captures the range of activities involved in conveying news and information to the public.

There is a fundamental difference between journalists working in news media, and those who work on features or documentaries. To a scientist, interacting with newspeople means the chance of exposure, perhaps only for a couple of seconds, to millions of people. But rarely is there the chance to offer more than a sound bite. News is immediate, and newspeople expect you to drop everything to talk to them. A deadline is a deadline. And even if you've spent an hour setting up for talking and filming with them for two minutes, there's no guarantee that an invasion or earthquake might not suddenly relegate your moment of fame to oblivion. Usually an interview takes the form of a phone call at an inopportune moment, with a reporter who hasn't researched any of the background to the story but has been told that something is exciting, and thinks you have nothing better to do than explain what it is and why it relates to life on Titan, or nuclear power in space, or whatever other agenda he or she is pursuing.

Documentary filming is generally more satisfying. The producers tend to have done their homework, and have an idea to whom they are talking and why. They might spend a day or two with you, or at least an afternoon. You get the chance to convey a real message (though you might be cut out nonetheless). And sometimes you get to do so in an interesting place.

"What bit of Earth looks like Titan?" I've been asked. "Well, who knows?" was the answer for many years. "We think it's an icy landscape, shaped by similar processes to Earth, perhaps with more impact craters. There should be lakes and seas, perhaps." At one conference in 2004 I remarked to journalists that Canada and Sweden both have fairly old rocks by Earth standards, and thus have comparatively many impact craters; they also happen to have lots of lakes. To my horror, this train of logic was abbreviated in the newspaper headline to "Saturn moon looks like Sweden," prompting thoughts of pine trees and blonde maidens. But really the question is as much one of "Which part of Earth looks most like Earth?" I had also advertised my vision of Titan as a southwestern desert analogue, cut by canyons and washes, a landscape shaped by flowing liquid, even though in general the land is dry. (This was, happily, one prediction I did get right.)

And so, back in August 2004, before the first *Cassini* flyby TA, I had found myself in Moab, in Utah's Canyonlands, blasting up and down a dirt road in a Jeep Wrangler, kicking up a trail of dust and splashing through streams. "Do it again, but try and get more splash," the cameraman had requested as he changed the filter on the camera. I readily obliged—it was jolly good fun. The BBC was filming a *Horizon* program on *Cassini* (also broadcast, renarrated, and edited in the United States as *NOVA*).

It was suggested by the producer that the opportunity for me to be filmed on a sky dive to parachute down into this landscape was available, to provide a visual analogue of the *Huygens* probe. Now, although I flew hang gliders as an undergraduate, the prospect of jumping out of a perfectly good airplane in the expectation that a bundle of cloth on my back would blossom into some descent-arresting structure seemed like an unnecessary leap of faith. In trying to understand the expected descent motions of *Huygens*, I had also had the opportunity to meet a number of parachute engineers (parachute aerodynamics is a particularly arcane field), and I found it striking that none of them dared try their creations themselves. If I was ever going to give skydiving a try and risk turning myself into a red smear on the desert, I'd do it after *Huygens*. A local parachutist was found to do the jump, a landlocked surfer dude who was psyched that someone would actually pay him to jump. He had a good time. I stayed unsmeared. Everyone was a winner. In the end, they used only a few seconds of footage from our two days, but it was a fun trip.

Now, post-*Huygens*, the BBC has persuaded me to be filmed again. This time it is doing a show all about Titan and wants to film terrestrial gullies just like the ones seen by *Huygens*. I make the familiar drive to Tucson airport, bright and early as usual. But this time, I turn off before reaching the terminal and go instead to a corner of the airfield where a helicopter company has been contracted by the BBC.

In fact, the helicopter and pilot are the same ones used for a calibration flight for DISR and the *Huygens* radar altimeter some years before. I gesticulate animatedly in the chopper at some vaguely Titan-like channels near Kitt Peak, and shout (in the hope that one or two will be usable) various permutations of remarks about how amazing it is that Earth and Titan's landscapes look the same, even though the working fluid and surface material are very different. We fly back to Tucson for some shots of me boldly striding to the helicopter, complete with close-ups of putting on my headset and buckling my harness. I am such a stud.

This time, the show uses the footage more extensively (the BBC could hardly fork out all that money on a helicopter and not use it). Viewers get, of course, a completely inaccurate view of how I spend my working day: much as I'd like to be driving Jeeps and buzzing around in helicopters every week, the reality is a little more prosaic. But that doesn't matter. The purpose of the show is to get some simple ideas across to people who do not have the luxury of exploring planets for a living, and perhaps have barely heard of Titan at all. And for that one needs better visuals than me typing e-mails in my office.

I remember my own teenage years: in 1986, I stayed up late to watch live coverage of ESA's *Giotto* probe encountering Halley—the first big European planetary success, featuring many scientists like John Zarnecki and Fritz Neubauer, who later were to go on to work on *Cassini*. And at age twenty in 1989, I was impressed as I watched a *Horizon* show "starring" Carolyn Porco and Larry Soderblom reporting on the *Voyager* encounter with Neptune and Triton. I vividly remember Larry firing off a carbon dioxide fire extinguisher to illustrate Triton's geysers. These people are now my colleagues, and I find that although Larry doesn't play with fire extinguishers all the time, he does interesting and important stuff nonetheless. Things have come full circle. Maybe there is some kid out there watching me who now thinks this exploring planets, driving Jeeps and helicopters business is a rather appealing one, and who will become a scientist or an engineer as a result.

......................................................................................................

## SWEATING RESULTS OUT OF THE DATA

In time, more and more scientific results emerged from the *Huygens* data. The temperature profile of an atmosphere is one of its most basic properties. The lower atmosphere profile (see chapter 2) was fairly well-known at one latitude, at least from the *Voyager* radio occultation, and the profile measured directly by *Huygens* under its parachute agreed very well with the (indirect) *Voyager* measurement. But at altitudes where the air was too thin for *Voyager* to measure, above about 180 km—all the way up to the ~ 1,000 km altitudes sampled by *Cassini* during its flybys, *Huygens* was able to measure the profile by recording the deceleration during entry— the air drag relates to the density. This analysis revealed remarkable fluctuations in temperature—similar to those seen in the stellar occultations in previous years. Titan's atmosphere was rich in structure—perhaps due to gravity waves or tidal effects. It was tempting to associate these fluctuations with the large variations in haze density seen in some *Cassini* images. But did the temperatures cause haze variations, or vice versa?

At the bottom of the atmosphere, the GCMS team had an interesting revelation. The methane abundance measured in their instrument in-

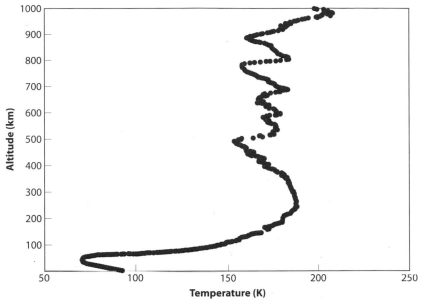

Figure 5.11. The temperature profile of Titan's atmosphere. Below 150 km the temperature was measured directly with thermometers. Above 150 km, the temperature is inferred from the deceleration of the probe—compare with the predicted profile in figure 2.07. The variable structures above 500 km are probably due to gravity waves. Compare also with images of Titan's haze in figure 4.11.

creased toward the surface, as one might expect it would by analogy with the water vapor profile in Earth's atmosphere. Evidently there was a source or reservoir of methane at or near the surface somewhere. But much more exciting than that, the methane reading took a dramatic jump after impact, suggesting that there was a reservoir or source of methane—perhaps as a liquid or as clathrate ice.

Even some of the simplest sensors can give interesting and important results. For example, the internal temperatures of the probe generally declined during the descent as ever-thicker cold air swept past and through it. Once on the ground, the cooling slowed down—and some components even began to warm up. It was possible to estimate the wind speed on the surface by how much windchill occurred. It seemed that the wind in the lowest meter or so of the atmosphere had to have been less than 0.25 m/s.

One heated component was the inlet pipe for the gas chromatograph/mass spectrometer. This was heated, albeit indirectly, to prevent droplets from blocking the pipe during descent. When switched on, the heater warmed up to about 90°C in the thin air of the stratosphere, but gradually cooled in the colder, denser air lower down. Then at impact, it warmed up again to 80°C, since cold air was no longer flowing through it. By carefully constructing a model of the heat flow from the heater to the inlet, it was possible to deduce that the ground was indeed damp with liquid methane (just as damp sand at the beach feels colder than dry sand); this meant that the methane did not come from some clathrate. But it wasn't just methane that was sweated out of the ground at the landing site. An unanticipated host of compounds, including ethane, carbon dioxide, and possibly benzene, was present.

Another constraint on the surface winds emerged. Careful study by Erich Karkoschka at the University of Arizona showed that the amount of light received by DISR's upward-looking photometer dipped slightly at impact. The dip was consistent with the parachute blocking off part of the bright sky as it fell to the ground. Karkoschka could even place limits on the wind speed.

Then there was the question of what exactly had happened at landing. In the impact acceleration record, there was some indication that the probe had bounced or slid somewhat for a second or two after impact. However, it didn't move far. Some of the optical data from the DISR camera showed that the scene in front of that part of the camera had not changed substantially. An interesting result emerged from the AGC analysis of the radio link. After the rapid modulation of the signal strength as the probe spun and swung its way down to the surface, the probe orientation remained fixed on the ground, and so the signal strength should have varied only gradually when *Cassini* set on the western horizon as seen from the probe. But in fact, the signal strength underwent several large dips that initially defied explanation.

What was happening was that, because the probe was sitting on smooth ground, the radio signal could also be reflected from the ground up to *Cassini*. At some angles, the direct signal and the reflected one added together, but at others the two waves were out of phase and canceled out, leading to large drops in the received signal strength. It was a textbook case of what communications engineers call multipath interference. (It is the same effect that leads to huge changes in cell phone signal if you only move slightly.) By fitting the detailed pattern of dips in the signal history, engineer Miguel Perez at ESTEC was able to deduce that the probe antenna was 75 cm above the surface or, in other words, that the probe was just sitting on the surface. Thus, it had skidded out of its hole or,

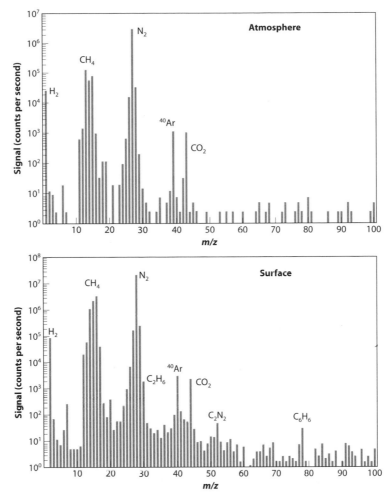

Figure 5.12.   The GCMS measured the composition of the atmosphere and, fortunately, the surface too. These are mass spectra; heavier molecules are to the right. Compare this with figure 4.08. The abundance of $CO_2$ was somewhat surprising. The argon abundance showed that Titan has been geologically active, and the detection of benzene showed that the surface is very rich in organic material. (Data courtesy of NASA GSFC/ESA)

rather, much of the 12 cm of penetration had taken place above ground, by driving some cobbles into the soft sand. This had left the probe resting on top without having itself penetrated deeply. It was even possible to show that, in the direction westward of the probe, where there were no good images, the roughness of the landscape was similar to that seen toward the south where DISR was pointing after impact.

Not all of the results made sense together—a symptom of *Huygens* being more heavily instrumented than *Galileo* or the *Pioneer Venus* probes. The SSP tilt sensors and a HASI accelerometer both seemed to indicate the probe was tilted by several degrees relative to the local vertical. But the horizon on the DISR images was pretty horizontal. It seemed unlikely that the ground would be sloping so steeply. Were some of the sensors wrong—and if they were, why did they agree with other factors? Maybe the probe had bent out of shape, changing the relative alignment. In fact, it would be impossible to know.

## LANDING SITE CORRELATION

The *Huygens* landing site location determined by the DTWG (Descent Trajectory Working Group) remained just a set of numbers for over eight months. Observations of the region by *Cassini*'s camera and the VIMS instrument did not show details of a scale that could be correlated with the DISR images. However, the ninth close flyby of Titan by *Cassini*, T8, on October 26, 2005, was to pass relatively close to the landing region, and included radar observations near its closest approach.

Although radar observations of the landing site were literally a long shot, during the planning stages a couple of years previously, it had been considered worth a try. Nominally, the radar was to work in its imaging mode only when closer than 4,000 km, but in principle, imaging could be done from farther away using at least the center beam, which had a stronger signal. Imaging the landing site this way would require a good estimate of where the probe was, since using only the central beam would give a very narrow view, like looking through a straw.

However, the first few flybys showed that Titan's surface was more reflective than had initially been feared. An echo strong enough for synthesizing images could be made from farther away than initially planned, and by setting the instrument parameters correctly, all five radar beams could be used to make a usefully wide image. It would still not be at such high resolution as the "real" opportunity to image the landing site on T41 in 2008, but it was worth a try while the *Huygens* results were still fresh.

..................................................................................................

RALPH'S LOG, OCTOBER 29, 2005

THE BREAKFAST CLUB

4:35 a.m.—my watch alarm goes off. The start of a long day—
and a Saturday at that. As we (the radar team) planned how
and when to work on the new data from T8, it transpired
that to get prompt results quickly enough for the various nine-
to-five public relations bureaucrats to approve press releases
in time for inclusion in Thursday newspaper science supple-
ments, *we* would have to work on a Saturday. It seemed somehow unfair, so I joked to the team that it seemed like some
sort of "detention" at school for bad behavior—reminiscent
of the movie *The Breakfast Club*.

Zibi gets up too, and before I am dressed has already logged
in to check the latest *Cassini* ISS images. Over her shoulder I
see some nice crescents—Titan images that I know must have
been taken as *Cassini* receded from Titan. "That's a good
sign"—if the images at the end of the Titan sequence are on
the ground, then chances are the radar data came down OK.
I gulp some orange juice and clamp a slice of toast between
my teeth as I head out the door.

Twenty minutes later, I park my Jeep at the airport. Ten
minutes after that, I am at the gate, well in time for my 6:00
a.m. flight. Tucson is a conveniently small airport—easy to
escape to or from. Soon after 9:00 a.m. I am in Burbank, California, and drive to JPL. Since it's not a weekday, the traffic
is light.

As I walk into the CSMAD lab at JPL (the Center for Space
Mission Analysis and Design, better known to us as the radar
"war room"), I see a radar image—just a small part, the high-
priority landing site region. It doesn't look like much, and
there will be more later, so I adjourn to a discussion on scatterometry and radiometry modeling, catching up on some is-
sues with measuring the noise floor in the instrument when

the attenuator settings are changed. Later, celestial mechanic Nathan Strange updates us on some of the planning activities for the extended mission. He shows us a very creative presentation of how *Cassini*'s orbit around Saturn maps into each Titan flyby geometry. But the elegant astrodynamics is swiftly abandoned when the news arrives that the full radar image has been processed—and it's a corker! But all we can do today is select a few choice pieces of the long image and draft press releases on how we see dunes and mountains. The data look great, but there is too much to digest and appreciate in a few short hours. Even though the real work of interpreting the data will take many weeks and hundreds of e-mails, the occasional face-to-face get-togethers are important for quickly exchanging ideas, especially when the team is confronted with exciting and puzzling new data. A few of us grab some dinner, and I head back to the airport. I arrive home close to midnight after my five-hundred-mile commute.

..............................................................................................

This radar sweep covered more terrain in one go than any of the previous passes—almost 2 percent of Titan's surface. And yet again, Titan looked different—no impact craters, and hardly any fluvial channels. But there were many of the dark stripes dubbed "cat scratches," like those seen in T3 in February (see chapter 6). However, many of these new ones were more than just dark stripes—they had bright highlights showing that they have positive relief. And they extended over a massive area, appearing to flow around mountains like the raked grooves in a Zen rock garden.

At the first viewing of the images, a number of vaguely circular features immediately excited the geologists on the team. They looked rather like volcanic calderas. And chains of mountains forming a strange chevron pattern were visible—a signature of tectonic stresses perhaps. Randy Kirk and Larry Soderblom of the U.S. Geological Survey, together with Lisa MacFarlane of the DISR team at the University of Arizona, tried to find a correlation between the radar image, which is most sensitive to slope and roughness, and the DISR images, which were sensitive to the brightness of whatever the surface is coated with. Some things looked as if they matched up, but it was much more challenging than expected: one match seemed as good as another. Matching the high-resolution optical map from DISR with a lower resolution regional optical map from ISS or VIMS and matching that with the regional-scale RADAR map seemed to help, but even then, there were a couple of equally persuasive matches.

A week or so later, however, after more considered study, a correlation emerged. The key was two dark streaks in the radar image that seemed to match up with two dark streaks in a few side-looking DISR images. The DISR team, shouldered with the responsibility of defining the real Titan and defending their conclusions to scientific peer review, were careful not to overinterpret horizontal stripes in data from the side-looking imager (SLI), which could easily be artifacts. Some amateurs, however, unburdened with such rigor, mosaicked the SLI images anyway, and their work showed a couple of dark stripes in the distance. Inspired with this clue, the SLI images were given a more thorough inspection, and Larry Soderblom nailed it.

Around five of the SLI images could be matched up with the dark radar streaks, and the high-resolution downward-looking frames taken at the same instant could be matched up with the rest of the DISR mosaic, making a geometrically accurate correlation between that and the radar image, which in turn was precisely tied to the orbiter's known trajectory.

Some other features in the mosaic matched up, but there were also some in the DISR mosaic that appeared not to have strong contrast in the radar image. The dark streaks were particularly intriguing, however, because these features were seen elsewhere in the radar images.

## RETROSPECTIVE ON *HUYGENS*

Even at the same time as analyzing the data and reconciling various discrepancies such as the apparently contradictory information from the pressure sensors and the altimeter, the instrument teams had one major duty, that of archiving the data. No one can anticipate all the useful analyses that can be made of data from planetary missions, and so it is important that it be placed in a public archive where anyone can access it. The same is done for Hubble Space Telescope images.

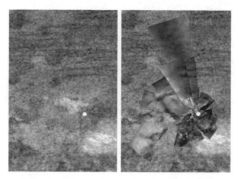

Figure 5.13.  A portion (about 80 km top to bottom) of the orbiter RADAR observation from October 2005. At right the mosaic of DISR images is superposed. The two dark sand dunes proved to be the key to identifying the landing site, shown with a small cross to the lower right of center. Otherwise, the small-scale optical and radar appearance of Titan's surface are surprisingly different. (Data: NASA/ESA/ASI/JPL/University of Arizona/USGS)

However, unlike HST images, which come from the same few instruments under much the same conditions, year after year, the results from a three-hour, one-off probe mission have all kinds of exceptions and complicating factors. These too have to be documented as best as possible—lest someone unfamiliar with the overall aspects of the project read too much into some rogue datapoint. Such documents are archived as plain text files, since it is not certain that any particular word processing programs will be available ten or fifty years from now.

One of the most remarkable products to come out of the DISR analysis was the painstaking work of Erich Karkoschka. Stitching together all the images and measuring their contrasts, he synthesized a movie of Titan as seen from the descending probe, complete with the sky color changing and the surface contrast improving at lower altitude. With such a product (he also made a geeky version with beeps and flashing lights to show when and where images were taken), it was easy to forget about the probe's strange spin and the missing data. Probably few people in the future will bother to delve back into the raw data because the top-level results—wind profiles, not Doppler histories; neat mosaics of images, not the raw, blurry thumbnails—are the *Huygens* legacy. But getting to those from the raw bytes that crossed a billion miles on January 14, 2005, was the product of an intensive year and one-half (and a particularly frenzied couple of days) of effort by a few dozen scientists.

The *Huygens* results set the stage for *Cassini*'s subsequent findings, of Titan as a world of truly Earth-like complexity. The picture of the

Figure 5.14. Synthesized views of Titan from the probe looking west (*left*) and north (*right*), seen from three different altitudes. The high-altitude views are blurred by the haze, but the two straight, dark sand dunes are visible north of the highland. At lower altitudes, progressively smaller features become visible. (ESA/NASA/University of Arizona)

"shoreline" with river valleys, and the pebble-strewn view of the landing site, would define this world for ever after.

But the mission was perhaps as important an engineering success as it was a scientific one. *Huygens* was, of course, a major coup for Europe, whose planetary science endeavors were ramping up (*Mars Express* in orbit, *Venus Express* under development, *Rosetta* on its way). Europe could put the *Beagle 2* disappointment behind it and look forward to future missions, and indeed *Huygens*'s success probably played a significant role in ESA's committing to future Mars landed missions through its new *Aurora* program. In the weeks and months after the encounter, SSP principal investigator John Zarnecki was visited by U.K. prime minister Tony Blair, and several French *Huygens* scientists were invited to the Elysée Palace to meet French president Jacques Chirac.

Psychologically, *Huygens* made Titan a real place—a place to which we might soon return. Before *Huygens*, Titan was a "here be dragons" sort of unknown place where one wouldn't dare venture. Now one could just show the *Huygens* pictures and the destination was clear. It was not much of an intellectual leap to imagine replacing *Huygens*'s battery with a long-lived power source and its parachute with a balloon; instead of exploring Titan for only a couple of hours, one could do it for years.

As engineers began to consider what new and wonderful machines might be sent to explore Titan in the future (see chapter 7), the phrase "Huygens results" would become a totem, some heaven-sent stone tablets of laws on the Titan environment, to which one had to defer.

Some results were truly surprising. The low abundance of organic gases in the lower atmosphere (such as ethane), compared with predictions, indicated that some loss process (or some detail of the chemical formation process) was not understood. And the presence of haze all the way down to the ground was not anticipated at all. Some other results confirmed expectations from models or *Voyager* data, but models can often be wrong. Only by making a direct measurement—by physically going there with an in situ instrument—could we be sure about the temperature profile, or the argon abundance and its implications for the origins of Titan's atmosphere.

Three decades of planetary science had previously had to make do with guesses and models, developed with much effort, that will now soon be forgotten. Those models had been necessary, indeed, to design the *Huygens* probe in the first place. But in future textbooks on planets—and despite being a satellite, Titan will doubtless deservedly appear in many (whatever the official definition of a planet)—the *Huygens* results will be taken as gospel, and quoted without much further comment as if they had always been known.

# 6. The Mission Goes On

Cassini's encounters with Titan occur generally at intervals that are integer multiples of Titan's orbital period of sixteen days. The gravity slingshots of the first encounters had brought *Cassini*'s orbital period down to twice that of Titan, and so the next encounter after the probe delivery on TC took place just a month after the *Huygens* descent.

Although in much of what follows, findings about specific locations were made with the VIMS and RADAR instruments, almost invariably where those locations were was substantially defined by the large-scale map generated from ISS images. As time went on, the map became progressively filled with more and more details, and more and more (unidentified) features, some with names even though it was not known what they were.

In this chapter, we describe events essentially in chronological order. It is important to remember that the findings on one flyby did not necessarily lead to any change in the plans of the next. *Cassini*'s operations are so complex and its pace of encounters so relentless that even minor changes need to be made weeks in advance, so that they can be checked. In a crisis, doubtless all the stops would be pulled out to fix an emergency, but given a finite budget and a four-year mission, things have to be done at an orderly, measured pace, in such a way that rework is minimized. So, most of the observations had been planned years before *Cassini* arrived.

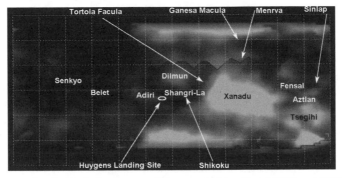

Figure 6.01. A cylindrical albedo map of Titan made from *Cassini*/ISS and HST images, centered on the anti-Saturn hemisphere. North is up. (Compare with figure 2.02.) Some of the growing list of named and known features are shown.

......................................................................................................

## RALPH'S LOG, MAY 2003

VENICE, ITALY

"Good morning!" A friendly British greeting flies from the vessel of one space explorer to another. It is Dr. John Zarnecki, the principal investigator of the Huygens surface science package, and Mark Leese, the program manager of that experiment. They are waving to me from the prow of a vaporetto waterbus, the path of which has just been crossed by my little traghetto boat. As the gondoliers row me across the canal, I turn and wave back to them, colleagues and friends for over a decade. My wave is a little unsteady because the boat is full: like the rest of the Venetian commuters in it, I am standing up. Will the *Huygens* probe rock in the waves like this on Titan? In less than two years, we'll find out.

We are in Venice for one of the thrice-yearly, weeklong *Cassini* Project Science Group meetings. Because of the substantial European involvement in the mission, and to share the joys and burdens of travel, one of these meetings each year is held in Europe, and so we are here, at the invitation of the Italian Space Agency (ASI).

There is, as always, concern about the science planning process, the massive effort to orchestrate *Cassini*'s activities for the four-year tour in minute-by-minute detail, before we actually get there. Not only does the pointing need to be planned, the choreography of the spacecraft turns, but also the data volume allocated to each observation, the amount of heating of sensitive instruments budgeted, transitions between operational modes decided, ground station availability checked, and so on. This is a huge effort, dubbed by some "Brian's Long March" (after Brian Pazckowski, leader of the science planning effort), and progress has been hampered by unforeseen subtleties in *Cassini*'s complex operations. But we are forging ahead and with luck should get most of the plan in place. Planning occurs in steps, with the level of detail and integration with other parts of the project increasing at each step. The number of sequences is such that planning must occur in parallel, with some sequences at step one while others are at more advanced stages. It would, of course, be the same individuals doing each—multitasking. The hope was that we'd get good at this quickly.

Another topic in the meeting, displayed on a computer projector set up in the church of San Lorenzo, causes some unease. One of the four reaction wheels that turn the *Cassini* spacecraft around has been stiffening up, friction in its bearings occasionally jumping to worrying levels. *Cassini* needs three of the wheels to fly normally, a fourth being carried as a spare. The spare is being switched on to replace the sticky wheel, so for now operations are not affected. But it is a reminder that, even though *Cassini*'s main mission has yet to begin, it is a veteran spacecraft and is starting to show its age. Like Hannibal's troops crossing the Alps before their battle with the Romans, our elephant-sized spacecraft has had an epic journey already. How will it hold up?

We sit in the church, the early arrivals among us as usual having strategically grabbed the seats by the wall in order to plug our laptops into power outlets for the long day ahead. I muse that the *Cassini* project has much in common with the religion for which our meeting room was originally built.

There is an entity somewhere in the sky, constructed by humans. Communication with this entity must be conducted through an elite priesthood, with appropriate incantations and an obscure vernacular (for example, one scripture, JPL document D-13242, *Cassini* document 699-063, the *Cassini* Project Acronym List, runs to thirty-six pages!). However, if we are good and hardworking now, there will be a fabulous reward in the future.

..........................................................................................................

## CRATERS, CHANNELS, AND CAT SCRATCHES

The T3 flyby on February 15, 2005, two Titan days after the *Huygens* arrival, brought a new radar swath and, happily, some geological features that actually made sense. One of these had been anticipated. On the TB flyby, ISS had seen a large, dark ring suggestive of an impact crater. The radar image left no doubt. It was a huge basin, measuring 440 km across the widest rings. The dark ring seen before was evident in the radar image as the rather flat floor of the basin. Perhaps it had been flooded with melt or magma released during the impact event itself (as with the basalt flooding of the lunar basins), or perhaps afterward it had been (or was even still) flooded with liquid hydrocarbons. Toward the center, a vaguely circular cluster of small, knobby hills stood above the floor, perhaps the vestige of a peak ring. The apparent rim of the crater was irregularly textured, with gullies cutting into the walls. In a lot of ways it resembled a somewhat smaller (280 km) crater, Mead, on Venus. It would pose a new challenge to modelers to see what the structure of this crater told us about the thickness of Titan's crust. Jonathan Lunine nicknamed it "Circus Maximus" as the radar team grappled with the image in the first few days, before it received its official name, "Menrva."

Curiously, the western side of Menrva was more heavily eroded and subdued than the eastern side. The same sort of east–west pattern had been noted in the ISS data, as if the motion of something from the west were doing most of the erosion and transport, or downhill were always eastward. Some scientists thought the planet was a puzzle like some Escher painting. Everything couldn't run downhill eastward all the time, could it?

Figure 6.02. Part of the T3 radar swath, showing the 440-km-wide impact structure Menrva. A small channel network is visible at lower left. Note the rubbly ring of hills in the center, surrounded by a smooth moat and a wide, gullied rim. North is at the top, and radar illumination is from the north. The swath is about 200 km across at this point. (NASA/JPL)

Breaking in through the southern wall of Menrva was a branched network of bright channels—rather evidently a river network. Had the *Huygens* images from the previous month not shown Titan to be a place modified by fluvial processes, that identification might have been more tentative. Channels are not always evidence of rain. For example, some striking, winding (albeit not as branched) channels were seen on Venus by *Magellan*'s radar, but these were cut by lava, not rain. Probably there was no liquid in the "rivers" on Titan, or they would more likely have looked black, but the bright appearance seemed consistent with what *Huygens* had seen. With cobbles 10 cm in size, a dry or damp riverbed would look rough, and thus bright, in the radar imaging.

More impressive was a network of river channels (again, downhill to the east) flowing northeast away from Menrva. Did that slope have something to do with Menrva, or was it a broader regional trend, a slope away from Xanadu? Did the rivers themselves have anything to do with Menrva? Some of the channels, which were generally narrower than the southern set, had wiggles and meanders, just like rivers on Earth. The longest extended for about 180 km, and the whole set seemed to drain into a broadly round, bright region. Interestingly, the channels branched and reconverged in places, a morphology termed braiding or anabranching. This is characteristic of rivers with lots of energy, that don't care if

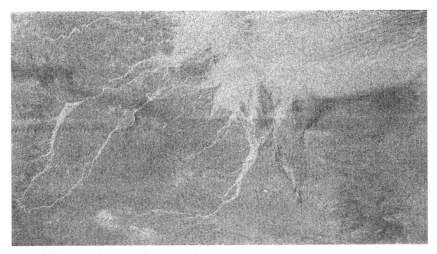

Figure 6.03.   Another part of the T3 radar swath, just to the east of that shown in figure 6.02, showing a braided channel network, perhaps characteristic of storm-generated washes or wadis. These channels are on a much larger scale than those observed by *Huygens* (figure 5.06). Again, the image is about 200 km top to bottom. (NASA/JPL)

there is a preexisting channel. They have enough energy to make a new one in the direction they happen to be going, whether there was an old channel or not. These tend to be seen on Earth in two environments. One is in deserts (such as the Southwest of the United States), where rain, when it comes, is in the form of violent thunderstorms. The other circumstance is somewhat peculiar to Iceland, where volcanoes can erupt beneath the ice sheet. When they do, water builds up under the ice until it breaks out in a sudden flood called a Jokülhlaup. Planetary scientists had become familiar with this phenomenon as an analogue to a similar event that may have occurred from time to time on Mars, on a much larger scale, to form a transient giant ocean, the Oceanus Borealis.

It was in the evening of February 15, after receipt by the Goldstone 70-m dish, that the data were passed on to the radar team to process into the image showing Menrva and the channels. Several hours later, an extra few minutes of radar data were processed. A glitch in the downlink had caused a small gap in the data, but beyond that was another part of the swath about 500 km long. Right in the middle of it was a perfect "little" crater. It was a respectable 80 km across. The geometric rigor of radar allowed Randy Kirk to deftly conclude that the structure, which had a remarkably flat floor, was 1,300 ± 200 m deep.

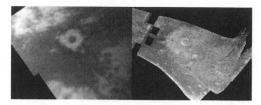

Figure 6.04.   Two views of the 80-km-diameter crater Sinlap. *Left*, a VIMS mosaic, showing a small bright area at the center of the dark crater, surrounded by a rhomboid bright area. *Right*, the RADAR image from T3, showing that the crater is flat-floored and some 1.3 km deep. (NASA/JPL/USGS)

This was quite shallow for a crater of this diameter on an icy moon. Furthermore, craters of this size tend to have central peaks or peak rings, yet this had nothing. Again, this suggested that the crater was degraded, perhaps filled in with sediment or at some time with liquid. In fact, it looked most like some Martian craters, many of which are similarly flat-floored.

There was also a rather puzzling, bright, parabolic region around the crater, probably an ejecta blanket, consisting of material thrown out of the crater when it was formed. In a later flyby, VIMS and ISS got a close look at the crater. In their lower-resolution data, it wasn't obviously a crater, but part of the ejecta blanket stood out. Indeed, this brought home the pitfalls of naming something before one understands it. The crater as identified by radar was named Sinlap, but the bright region around it, which had been identified as a "bright region" by ISS, had already been given the name Bazaruto Facula, *facula* being the term for bright regions (*macula* the one for dark regions).

Intriguingly, the floor of the crater was optically dark, as with Menrva, and the VIMS image showed a small bright spot at the center that didn't have an obvious RADAR counterpart. Clearly the optical and radar instruments were showing us different aspects of Titan's surface. It would take a while to fully understand the integrated picture.

One of the most prominent features across the T3 swath was difficult to interpret. Patches of dark streaks appeared from place to place, generally oriented roughly east–west. In some cases, they curved or merged. No topography was obvious, and one idea was that they were thin streaks of material on the surface. Another was that they were some kind of seep of liquid. Because they looked like claw marks, the radar team nicknamed them "cat scratches." (For some reason, *Cassini* scientists seemed

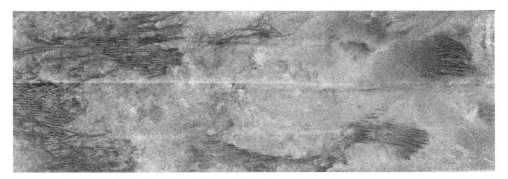

Figure 6.05.   The strange, dark, quasi-linear features seen in T3 in February 2005 and nick-named "cat scratches" were suspected of being aeolian features. Their nature became clear on T8 (see figures 6.08–6.10). (NASA/JPL)

to have a preoccupation with cats. As well as cat scratches, they were talking about tiger stripes on Enceladus, and Si-Si the Halloween Cat.)

Although one area of the T3 swath looked reassuringly similar to TA, which is to say somewhat jumbled, in general it almost seemed that TA and T3 had looked at different planetary bodies—which was good; a diverse surface is a more interesting one. Disappointingly for the radar team, which had as many vulcanologists in it as there were candidate volcanoes on Titan, there were no obvious volcanic features visible on T3.

## THE BRIGHTEST SPOT ON TITAN

In the southern part of Xanadu, at 80° west and 20° south, the VIMS team noticed a bright oval area. It wasn't especially bright at the wave-lengths (e.g., two microns) where good maps had already been made, but at five microns, the longest window available to VIMS, it stood out from its surroundings in a striking manner.

This excited the team. Could it be bright because it was glowing "hot"? The fact that it was much brighter at five microns than at shorter wavelengths was consistent with it being due to thermal emission from, say, a cryovolcanic flow with a temperature a hundred degrees above the rest of Titan. That would be an exciting discovery indeed, but the acid test would be to observe the area at night, and an opportunity to do that was not immediately expected.

Figure 6.06.  A VIMS mosaic showing the Fensal-Aztlan ("H") region. Sinlap is visible just to the right of center. A strikingly bright area (especially at five microns) is seen at left, between Xanadu and Tsegihi. (NASA/JPL/University of Arizona)

But other data shot down that idea anyway. *Cassini*'s microwave radiometer had observed the region, albeit obliquely, and had detected no enhanced thermal emission. Presumably, then, the feature was related to a surface deposit of some distinct material. But it wasn't fresh water ice, washed clean perhaps by a methane rainstorm. Water ice is dark at five microns. The most obvious simple material was carbon dioxide ice—but this was not considered the most probable substance to be associated with resurfacing. The prevalent idea was that any cryovolcanic liquid on Titan would most likely be a water–ammonia mixture, since ammonia can lower the freezing point of water to only 176 K. Ammonia also has the advantage of reducing the density of the cryomagma. Without it, water is rather more dense than ice, and so would be unlikely to ascend from the interior.

Although the oval patch itself was not especially bright at 0.94 microns, the surface map made by the ISS team did show that its southern margin ("the Smile") was bright. Although this was an interesting feature, what it meant was far from obvious. At the very least, the feature deserved to be flagged as an important site for further observations, such as radar imaging.

It seemed that trying to find active volcanism on Titan had become a consuming passion for the VIMS team. Bob Nelson at JPL had monitored several spots on Titan with VIMS data and claimed that one of them changed character from one flyby to another—perhaps a cryovolcano. But extraordinary claims require extraordinary evidence, as Carl Sagan once said. Most scientists in the community argued that the changes might be due to different viewing geometry, that differences in reflectivity at different angles or a different path through the hazy atmosphere was the reason it looked different. Nelson maintained that these effects had been taken into account in his analysis, but many seemed unconvinced. Maybe he'll yet be proved right.

Making the right judgment about when to go public with a finding is tough. As the months and successive flybys follow each other, the evidence for or against would add up and the case would become more solid, but wait too long and one might get scooped by another instrument or even a different faction on the same team. Is the person who perceives a pattern emerging before anyone else a visionary, or just seeing imaginary things? Thus, there is an art to doing science. The VIMS team kept looking for changes—although a change in surface coating needn't be attributed to volcanism necessarily; the removal of sediments by wind or fluvial action is a less extraordinary process that might also change the surface.

## CLOUDS COMING AND GOING

Titan's polar regions seem to be particularly dynamic. Over the winter (north) pole, the haze patterns continued to evolve, with the extended polar hood somehow connected to the detached haze layer. And as discussed in chapter 4, the bright and persistent clouds that had surrounded Titan's south pole during *Cassini*'s approach and early flybys seemed to have disappeared by early 2005.

However, ground-based observers using the Keck 10-m and Gemini 8-m telescopes with adaptive optics, including Henry Roe and colleagues at Caltech, had noticed an intriguing trend in the clouds at 40° south. These tended to last only one terrestrial day in contrast to the polar clouds, which often lasted for weeks. Moreover, they tended to appear at the same longitude—about 350° west. These clouds could have been present before 2003, but there had been too few observations at that longitude to tell.

The preferred explanation put forward by the observers was that there might be a source of methane at this location—if not a full-scale cryovolcano, then a vent or geyser of methane. This, at least, would explain the

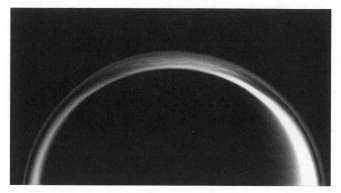

Figure 6.07.  An ISS image of the north polar haze taken in January 2006 from some 2 million km away. The complex structure of the polar hood is evident, as is the detached haze extending south. Compare with the *Voyager* view in figure 2.11. (NASA/JPL/Space Science Institute)

preferred location, and it was this suggestion that got the headlines when their findings were published.

Perhaps a more prosaic explanation was that a mountain caused clouds to form—orographic clouds. As air is pushed upslope, it cools, causing condensation. However, it was certainly an area that deserved a closer look, and the episode confirmed that, even though *Cassini* was at Saturn, ground-based telescopes were still making an important contribution.

Meanwhile, analysis of VIMS data from earlier flybys was coming along. The data included observations of some of Titan's clouds between 40 and 60° south during the TB encounter, although at different longitudes. The VIMS spectral cubes showed the outline and location of the clouds at hundreds of wavelengths, and by modeling the spectrum, Caitlin Griffith and colleagues were able to infer the altitude of the cloud top and the thickness of the clouds. Image cubes taken some tens of minutes apart showed striking changes in the altitude of the cloud. The cloud top of one blossomed up from 20 km to 42 km in thirty-five minutes. In other words, although cloud convection on Titan is rare, it is not much less vigorous than on Earth, and we could see the process in action!

Griffith and colleagues preferred a meteorological explanation for these clouds, which at least behaved like convective clouds on Earth. They argued that 40° south would be where upwelling motions would be expected in late summer. Nevertheless, they acknowledged that the apparent preference for some longitudes did implicate the surface in some way.

## NO SPECULAR GLINTS

Given the expectation of finding liquids on Titan's surface, and in particular finding them in optically dark regions, the lack of a specular glint was a persistent mystery. Even from Earth, glints should have been visible, and observers at the Keck telescopes were puzzled when they did not find them. However, this didn't mean that Titan was dry. All it meant was that there wasn't any flat liquid at a number of points along one line of latitude, basically between the sub-Earth point and the very nearby subsolar point. Perhaps that latitude was dry, or perhaps it was windy enough to generate waves.

But as time went on, the number of places where a specular glint might have appeared, but did not, grew. In particular, once *Cassini* had arrived with its ever-changing viewpoint over Titan, the range of potential specular points increased dramatically, but still there was no glint. Bob West at JPL and his colleagues evaluated how much the haze should suppress the glint, but even with the haze taken into account, it should have been visible. And so the conclusion was clear, as betrayed by the title of their paper: "No Oceans on Titan from the Absence of a Near-Infrared Specular Reflection." They reconciled the apparent radar specular glint with the lack of an optical one by suggesting that liquid had been present at some time, creating flat surfaces, but was no longer exposed. A natural analogue might be a dried-up lake bed or playa, or the enormous flat expanses of the Martian northern lowlands, the "Vastitas Borealis"—perhaps the bed of the sometime putative "Oceanus Borealis." (The topic of disappearing oceans is a recurring one in planetary sciences.)

## A POOL NEAR THE POLE?

Although there was little or no evidence of standing bodies of liquid at low latitudes, a distant flyby by *Cassini* in June 2005 gave it a good, if far-off, view of the southern polar region—by then free of clouds. The ISS snapped a rather appealing dark spot, looking for all the world like a backyard swimming pool or a kidney bean—or like a lake. In fact, the feature was very similar in size to Lake Ontario on Earth.

Of course, the fact that it was lake-shaped and dark didn't make it a definite lake, any more than the croissant-shaped radar-dark features

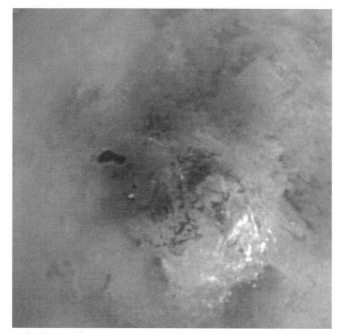

Figure 6.08. An ISS image of the south pole taken in June 2005 from 450,000 km away. Most of the clouds have disappeared, compared with the T0 images (figures 4.01 and 4.02) taken a year previously, though a few stragglers remain at lower right. Just left of center is a kidney-shaped dark feature, which could be a lake. (NASA/JPL/Space Science Institute)

seen by radar on TA were lakes. Maybe it was just a lake-shaped patch of soot, but it was the best candidate yet. Unfortunately, these high southern latitude locations would not be examined closely until late in *Cassini*'s tour.

In the meantime, in late spring 2005, the pace of *Cassini* flybys slowed somewhat, giving many teams a chance to catch up with analyzing their data. Flybys T4, T5, and T6 added to the inventory of optical and particles-and-fields data, and measurements of the upper atmosphere showed a definite gradient in atmospheric density with latitude. This was important because it had become clear that the altitude of 950 km that the project had planned for its lowest flybys would have a higher density than had been predicted. This was, after all, the reason for the prudent strategy of lowering the flybys slowly at the beginning, testing the water as it were.

## LUCKY 7?

Many Titan findings were first reported widely at the 2005 Division for Planetary Sciences (DPS) meeting, in Cambridge, United Kingdom, in the first week of September. During the five-day meeting, a whole day was devoted to special *Cassini* talks, and more than another half day was devoted to Titan-specific talks. The *Cassini* project held its Project Science Group meeting in London the week before, making it a combined trip.

The T7 flyby on September 7, 2005, was an important one for RADAR—it would be the lowest flyby so far, and would sweep out a great arc on Titan, from a little east of Xanadu down almost over the south pole and back up on the anti-Saturn side of Titan. Its timing was a little awkward, occurring just at the end of the DPS meeting. Although some of the radar team gathered at JPL to study the new data as soon as it came in, the DPS meeting meant others had to teleconference in from the United Kingdom.

But, as they dialed in late in the U.K. evening, they learned that something had gone wrong. There was a bug in the software controlling *Cassini*'s solid-state data recorders. These 4-GB data recorders were the latest technology when *Cassini* was designed, and much more reliable than the tape recorders used on spacecraft like *Galileo* and *Voyager*. Nowadays you can put this amount of storage on a key chain in your pocket, although it wouldn't be resistant to space radiation. The software bug caused the recorder to think that it had run out of space about halfway through the flyby. As a result, it failed to record any more data. Worse yet, an entirely separate problem on the ground caused one of the Deep Space Network antennas receiving the data from *Cassini* to be pointed temporarily at the wrong part of the sky.

The problems hit all the teams hard and in more or less equal measure. All the outbound observations from the Titan pass were lost, including the second half of the radar swath. But what was received was exciting. Once again, Titan looked like a bizarre new place. A complex of hills was evident at the northernmost start of the swath, grading into a highly dissected area. Soon thereafter, several river channels appeared, flowing south. And then, just before the swath was truncated by the recorder problem, an irregular scalloped boundary appeared, beyond which Titan

Figure 6.09.   The end of the T7 radar swath. Right is toward the south, ending about 75°
south. Top to bottom is about 200 km. The scalloped bright–dark margin seemed suspiciously
like a shoreline, with the near-featureless dark area to the right either a hydrocarbon sea or at
least a smooth seabed. (NASA/JPL)

seemed dark and almost featureless. Was this a long-sought coastline—
was that the sea? One couldn't be sure. Could the faint textures in the
dark plain be changes in sea state due to winds, or was it the muted
signature of the seabed seen through a shallow radar-transparent layer
of hydrocarbon liquid? Or was it a relic, a coastline marking the edge of
a sea that was no longer there?

## WHAT IS THE SURFACE REALLY LIKE?

The nature of Titan's surface actually at the *Huygens* landing site seemed
pretty clear. But beyond the fact that it was at the edge of an optically
bright region, not much was known about it from orbiter observations.
It would be another few months before a specially designed radar observation would catch the landing site on T8. And that didn't necessarily
mean that the rest of Titan's surface, for which we wouldn't have in situ
data, would make sense. In particular, the nature of the large bright and
dark regions on Titan's surface were still basically unknown.

A Titan Surface Workshop in the pleasant summer of northern Arizona in July 2005 had taken place at the U.S. Geological Survey in Flagstaff and had given the opportunity to review what was known, in particular about Xanadu. Several other such informal workshops, of about twenty people actually working on the data, took place at regular intervals, giving the *Cassini* teams the chance to compare (and criticize) early results before they were exposed to a wider audience at big conferences like the DPS.

Of course, when Xanadu had been first discovered by HST in 1994, there were speculations about what it might be. The usual explanations were invoked—a volcano or an impact crater—although there was no particular reason to suggest either of these. It was just a big bright patch. A favored interpretation, on aesthetic grounds as much as anything else, was that Xanadu might be highland terrain, washed clean of dark organics by methane rainfall. Since that time not much more had been learned about it, except that its near-IR spectrum seemed incompatible with pure water ice, and that it seemed radar-bright compared with other areas, as well as optically bright.

Xanadu is on Titan's leading edge, centered at about 90° west, and *Cassini*'s flybys almost invariably have the spacecraft hanging above 0° and 180° on its way into or away from the inner Saturnian system. So there was only a limited amount of information about Xanadu, mostly on the edges, even at this stage.

Bonnie Buratti on the VIMS team presented an analysis of the phase function of Titan's surface—that is, how its near-infrared reflectivity varied with viewing angle. This analysis seemed to suggest that Xanadu was rough, at least on a very small scale. Xanadu had been observed with the radar, in its low-resolution scatterometry mode. This mode can operate from farther away than the imaging mode but yields only very low-resolution maps of the radar reflectivity. Nevertheless, it showed that Xanadu was much more radar-reflective than the dark areas to the west. There were several possible interpretations. It could be made of intrinsically more radar-reflective stuff, perhaps more ice-rich than organic-rich. It might be rougher at the radar wavelength scale (2.2 cm). Alternatively, there could be some exotic subsurface scattering mechanism that made the region backscatter strongly, like the "cats' eyes" in a road.

Was Xanadu just a giant version of the highlands to the north of the *Huygens* landing site? Or was it somehow mantled with some deposit of frost, the cryovolcanic equivalent of ash? Time would tell. At least both the optical and radar stories seemed somewhat consistent. But just as mysterious were the large dark areas seen all over Titan. Were they all dark in the same way? Were they just like the pebbled dark plains on which the probe had landed? A partial answer, and a surprising one, came in October 2005.

## THE SAND SEAS OF TITAN

Even in the earliest HST maps of Titan, observers noted that the center of Titan's trailing hemisphere was dark. The speculation at the time was that there might be lakes or seas of liquid hydrocarbons. This simply followed the tradition that seas are dark, like the "Wine-Dark Sea" in Homer's *Odyssey*. The historic interpretation and naming of the dark lunar plains as seas (*maria* in Latin) is a case in point. So, dark things are seas.

The high-quality maps generated in 2003 by the Keck telescope (see chapter 2) and by *Cassini* in 2004 showed clearly what had been only hinted at before—namely, that the darkest regions seemed to be concentrated around the equator. However, no specular glint, the mirrorlike reflection of the Sun on a smooth surface, even a dark one, had been observed, and this posed a challenge to the "obvious" interpretation as seas.

The T8 radar observation solved the conundrum. It was known already that on the large scale, the equatorial dark regions were somewhat radar-dark as well. Furthermore, the T8 radar swath had dark "cat scratches," like a few areas of the T3 one, but it also had some examples that showed relief. They were evidently tall enough for the radar beam to glint off slopes oriented toward the spacecraft. They were long, narrow mounds, not just streaks painted on the surface. They were sand dunes— giant ones, a kilometer or two wide and a few kilometers apart, but tens of kilometers long.

Sand dunes come in a variety of shapes and sizes. When wind blows over a uniform bed of sand, the sand particles begin to move by rolling and bouncing along as the wind picks up speed. Once they start to move, the impact of some grains helps to get others going, and a layer of bouncing particles builds up. You can sometimes see this at the beach or in the desert. The process is called "saltation," from the Latin word for

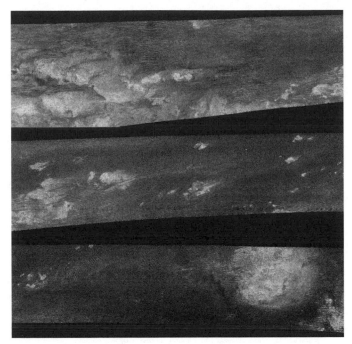

Figure 6.10. The T8 radar swath was much longer than prior observations, exploiting the RADAR instrument's performance to the full. Vast areas proved to be covered in dark material, punctuated by a few bright hills. The dark material appeared organized in near-parallel streaks, interpreted to be linear sand dunes. The swath shown varies from a little over 150 km wide to over 200 km. The dunes stretch for over 1,000 km. (NASA/JPL)

"jump." It occurs at wind speeds lower than those needed to lift the particles up altogether and suspend them as in a dust storm, so saltation happens more often.

But the sheet of sand doesn't move uniformly, and in time, beautiful structures emerge. Grains will pile up against a random obstacle, forming a ramp that traps other grains, and so a pile or a wall of sand builds up. But as the pile grows, it increases the wind speed above it, and so the sand is moved most quickly from the top of the pile. This sand removal prevents the pile from growing indefinitely. Consequently, piles or walls tend to have a characteristic size. The emergent pattern of sand piles depends on the wind regime and on the availability of sand. In a uniform wind, with plenty of sand, "conventional" transverse dunes, like giant ripples, form orthogonal to the wind direction and slowly march downwind. If the region is starved of sand, isolated crescent-shaped "barchan" dunes form.

Figure 6.11.   Close-up of some regions of T8, showing the topographic glints that revealed the streaks to be positive relief-dunes. The interaction with bright hills, parting and reconverging, showed them to be linear or longitudinal dunes, indicating the average wind pattern. Radar illumination is from the north, at the top. (NASA/JPL/University of Arizona)

If winds are highly variable, star-shaped dunes can form. But one of the most common forms on Earth is the longitudinal dune. This type of dune arises when the wind tends to fluctuate about a mean direction. As sand is pushed one way and then the other, it tends to accumulate in a long, narrow mound that lies along the mean wind direction. Such longitudinal dunes are sometimes also referred to as "seif," the Arabic word for "sword," though this term strictly applies to curved longitudinal dunes often formed by merging barchans. Longitudinal dunes are rare in the Americas, but are very common in the Namibian desert and in the Sahara, Arabian, and Australian deserts.

It was obvious from the T8 radar data that the Titan dunes visible were longitudinal. Not only did they just "look" longitudinal in terms of their shape and arrangement, sometimes joining to form "tuning fork" junctions, but they flowed around mountains and other preexisting topographic obstacles. The effect was to make this area of Titan look like a Zen rock garden, with grooves raked around rocks standing like islands in a sea of gravel.

Figure 6.12.   A handheld digital camera image of the west coast of Namibia, taken by an astronaut on the orbiting space shuttle and showing the giant longitudinal dunes there, which are exactly the same form and scale as those seen on Titan. The South Atlantic is at the top of the frame. (NASA/JSC/EOL)

The presence of these dunes immediately meant a number of things. First, there was an abundance of sand. Knowing the size of the dune fields to be about 200 × 2,500 km, with dunes around 100 m tall, meant there were tens of thousands of cubic kilometers of sand. The term "sand" in the context of Titan refers to the size and nature of the particles, and of course does not imply it has a similar composition to the sands of Earth or Mars. Whatever the sand was made of, it was optically dark, suggesting perhaps the presence of organic materials rather than ice, and it was not sticky. In other words it was dry, at least sometimes.

One of the most striking fields of these longitudinal dunes on Earth is found near the coast of Namibia. These dunes can be over a hundred meters high, and extend in long lines running roughly north–south, sculpted by winds that change direction with the seasons. They are impressive even seen from space, and inspired space shuttle astronaut Story Musgrave to write these lines:

> Now, Namibia, desert streaming into ocean,
> waves of bright sand diving into dunes of dark water
> -visible rhythms of blue and brown,
> sea and sand dance upon my strings.

In January 2006, Pascal Rannou and colleagues in Paris published the latest results of simulations of Titan's wind and clouds. Their computer model showed that there were several different categories of cloud—not only the large but occasional features seen near the pole where it was summer, but also activity in mid-latitudes in spring and fall. It even "predicted" clouds at 40° north in the 1995–97 time frame, when a large cloud had been tentatively discovered there in Hubble Space Telescope data. The model also suggested that low latitudes would see very little rain. This seemed to fit with the sand "seas." Slowly, a coherent picture of how Titan works was starting to emerge.

## TITAN—IN "TECHNO-COLOR"

Sometimes extracting new results doesn't require new data so much as new blood. The established scientists on *Cassini*, like any others, all have their favorite problems and approaches. Perhaps this was even more noticeable than usual because it had been so long since the teams were originally selected. Tried-and-tested methods that yield robust results when confronted with a few hard-won spectra are all very well, for example, but the VIMS instrument was returning literally thousands of spectra at every flyby.

A young computer-savvy postdoc, Jason Barnes, decided to challenge this mountain of data by automatically classifying spectra into a tractable handful of categories. His initial approaches with data through T6 were promising, but his PC quickly ran out of memory. Within a few weeks, with a bigger PC and more efficient code, and yet more data, he had made a near-global map. But a map of what? The code could be tuned

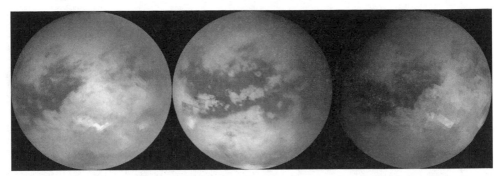

Figure 6.13. A VIMS map reprojected onto a sphere viewed from three longitudes. (Compare with figure 6.01.) Some prominent color differences are seen (see color insert). (NASA/JPL/University of Arizona)

to classify the spectra—hundreds of wavelengths for each pixel on the map—into as many types as the observer could handle without his or her brain exploding, and display different classes as different colors. It was an impressive map, brimming with information, and since color images are particularly sought after by the media, Barnes's product received a lot of attention.

This classification process said nothing about what each color meant. But that didn't matter. Someone else could deal with that. It was a nontrivial exercise for Titan, since the pesky atmosphere kept getting in the way and didn't even have the decency to stay the same everywhere on Titan or even as the mission went on. Perhaps one of the "old guard" would fit the spectra, or maybe some fresh new victim on the *Cassini* treadmill would do it. Barnes's approach got the process going, making it easy to identify which areas on Titan were different, which the same, and which, perhaps, were interesting enough to explore further.

The period between T8 (October 2005) and T15 (July 2006) was also one when *Cassini* was spending its time in the ring plane, which led to many encounters with Saturn's other satellites, as well as some spectacular juxtapositions of Titan against the rings, or other satellites. High phase-angle images were taken of Titan's crescent (or ring), which showed changes in the haze distribution. The haze over the pole was strikingly, and unaccountably, collected into mini-layers, an effect some scientists attributed to gravity waves.

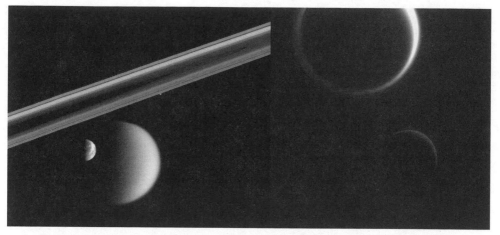

Figure 6.14. The low-inclination period from T8 through T15 kept *Cassini* in the ring plane, making for many spectacular juxtapositions of Titan, the other moons, and the rings. At left is a half-full Titan, with Dione in the foreground to its left and the rings (and little Prometheus) above. The right panel is a high-phase Titan, with a high-phase Tethys in the foreground to its lower right. The Tethys crescent shows the direction to the Sun, but unlike Titan's hazy ring, its crescent goes only halfway around the limb. (NASA/JPL/Space Science Institute)

## SPOUTING ON ENCELADUS

Among the most anticipated of the satellite encounters was one with Enceladus on November 29, 2005. Enceladus, tiny as it is, had long been suspected of having volcanic activity of some kind. Its surface was fresh and smooth in places, and the faint E ring of Saturn was brightest just at the distance of Enceladus, as if the moon were the source of the ring material.

A flyby in February 2005 had given a tantalizing hint, albeit an indirect one. Specifically, the magnetometer team had noted that Saturn's magnetic field was draped around Enceladus as if something bigger than the moon itself were getting in the way of the plasma flow that is dragged by the field. It seemed to be the signature of an atmosphere.

The *Cassini* imaging team had seen, in January and February, some faint striations suggestive of a plume. But *Cassini*'s imager, like any other camera, could be fooled by scattered light if a bright object is in or near its field of view. Some scientists who saw the images were pretty sure they were seeing a plume of material, but others were cautious. Unless there was solid evidence to back up the assertion, it would do nobody

Figure 6.15.  More juxtapositions. *Clockwise from top left*, Titan seen across the rings with Epimetheus just above; high-phase Titan seen through the rings; Titan's haze, this time in the foreground, with the stripy shadowed north pole of Saturn peeking through above; high-phase Titan seen across the rings and Saturn's crescent limb. (NASA/JPL/Space Science Institute)

any good to make such a significant claim—volcanic activity observed in action for the first time in the Saturnian system, with the implication that there was liquid water inside a moon that was by rights too small to have any such activity. The team resolved to make more robust observations at a future flyby, taking images with the camera (which is to say, the whole spacecraft) in different orientations. If scattered light were causing the appearance of a plume, then it would be in the same place on the images as the camera was rotated, whereas a plume on Enceladus would stay fixed relative to the moon. This was a very sound approach. When the images came back, however, caution seemed no longer necessary. There was little doubt what they showed. Enceladus looked like a comet, with a set of finely collimated streaks of brightness jetting out from around its south pole.

The images from the November flyby made data from other instruments, on a very close encounter in July, seem much more real. On that occasion, motivated by the tentative detection by the magnetometer, *Cassini* was guided low over Enceladus's south pole, only about 168 km above the surface. *Cassini*'s UVIS instrument, by studying the occultation of the star Bellatrix (Gamma Orionis), confirmed the existence of the plume and enabled the measurement of the amount of gas as a function of altitude. Absorption due to the vapor was apparent only on one side of Enceladus, on the ingress of the occultation. On the egress side, the light level jumped straight up as the star emerged from behind Enceladus's limb, with no slow buildup due to decreasing absorption as the plume thinned into space. The same instrument had made occultation measurements during the February flyby, but without detecting anything; so the plume was an elusive beast, appearing only in one out of four occultations. A remarkable finding was that the amount of vapor implied by the observation of the plume could supply all the water needed for the faint cloud of water-derived gas that surrounds Saturn.

*Cassini*'s dust counter, for which the Enceladus encounter was arguably the mission highlight, registered a flurry of impacts on *Cassini* as it flew through the plume. The peak in dust count on the 50 $cm^2$ impact detector was about four impacts per second, occurring about one minute before the closest approach of *Cassini* to Enceladus.

The INMS instrument measured a peak gas density a little later than the dust peak, almost when *Cassini* was closest to the plume, and confirmed that most of the plume material was water vapor. Surprisingly, virtually no ammonia was present. So much for the old ideas about ammonia depressing the freezing point. A couple of percent of nitrogen and a few percent of $CO_2$ were detected. Compositionally, Enceladus seemed not unlike a comet.

The velocity with which the plume emerged gave some clues as to the nature of the eruption. The plume dynamics revealed by the images required that the dust be accelerated to several tens of meters per second, which required the gas that drags the dust out to be moving much faster, and that, in turn, meant that the erupting fluid had to be warmer than 250 K. It argued against a cold ammonia—water magma and rather for straightforward liquid water, perhaps spritzed up with a little $CO_2$. *Cassini*'s CIRS instrument measured some surface temperatures as up to

Figure 6.16.   The fountains of Paradise? The crescent of tiny (250 km) Enceladus is adorned with finely collimated jets, like those from a comet. The narrowness of the jets and their height suggest that liquid water is close to the surface. (NASA/JPL/Space Science Institute)

157 K, considerably warmer than the Sun's radiation alone could cause the surface to be.

The question arose of what to call these Enceladus features. The material in them was water ice for the most part, but a geyser is a rather particular sort of thing—typically periodic and controlled by the weight of a column of liquid. Calling them volcanoes seemed a bit strong—they weren't really built up, and at a couple of hundred kelvin, they weren't exactly red-hot. Jets—like those of a comet—or plumes seemed to be the most accurate and descriptive terms.

The activity on Enceladus had no real physical effects on Titan, but it did have implications for Titan exploration. The thinly stretched scientific workforce on *Cassini* was not unreasonably compelled to devote its attention to documenting Enceladus. And it became impossible to ignore Enceladus as a possible focus for *Cassini*'s extended mission, or even a future mission. Titan might have to share.

The findings from any one of these instruments would have been interesting and exciting. But this encounter demonstrated the power of *Cassini*'s payload. Only by bringing the data together could a convincing and coherent impression of this remarkable phenomenon be gained. The same would be true of Titan, but the sheer complexity of that world would defy quick and easy analysis, and the relentless pace of *Cassini*'s Titan encounters would keep the Titanophiles too busy to contemplate their data for long. This underscores the importance of the public archive, the mechanism by which the data are made available to everyone (after checking for errors). The existence of the archive means that the wider scientific community can apply new methods and approaches to analyzing it.

The activity on Enceladus set the stage for contemplating cryovolcanism on Titan. Some high-resolution VIMS data of part of Xanadu on Titan showed what could be lava flows on the same scale as the feature seen by RADAR on TA. These exhibited a sequence of brightness changes, as if flows were progressively discolored by dust deposition or some other aging process. A similar trend in the cracks on Europa had been used to infer an age progression, which paralleled an orientation change, suggesting that Europa's crust had rotated relative to Jupiter. And there was the question of what made the five-micron bright spots on Titan. If Enceladus were erupting $CO_2$, why not Titan? $CO_2$ ice crystals, freezing out of a vapor plume, could be making the regions Hotei and Tui, in the southern part of Xanadu, bright.

## THE TOUR GOES ON

The *Cassini* project had always been designed with a "nominal" mission duration of four years, to end in June 2008. A nominal mission is the benchmark to which all the initial budgets are planned, and not achieving it would be regarded as a failure of some sort. But, as is often the case, a combination of good luck, conservative design, and robust margins means that the capability to continue operating beyond that nominal mission exists. Such extended missions are a bargain. The data return from a mission can be doubled or more for only the modest cost of continued staffing for operations. There is no new spacecraft to build. *Galileo*'s mission at Jupiter was extended several times (from a nominal mission of two years to a total of eight), and the *Mars Global Surveyor* (MGS), which arrived at that world in 1997, is still operating as we write in summer 2006.[1]

---

[1] In fact, contact was lost with MGS at the end of 2006, as this book was entering production.

*Cassini*, and its fuel budget, were holding up well, and the opportunity to make more measurements with its arsenal of instruments in an extended mission was something that the *Cassini* scientists were beginning to bank on. But one cannot wait until the end of the nominal or primary mission to start thinking about what to do. It would take years to plan.

And so, fall 2005 marked the commencement of planning in earnest of the extended mission. Even while grappling with the day-to-day running of the spacecraft, and struggling to keep up with the deluge of data, the science teams had now to engage in a new job. What should be the emphasis during an additional two years in Saturn orbit beyond the primary mission? As ever, different scientists had different preferences. Although Titan was obviously an important target, Enceladus had emerged as also deserving attention. And so, new tours would be generated and evaluated. One clever idea, being studied as this book nears completion, is ending the tour in a resonant so-called cycler orbit. This orbit, once attained, would need very little fuel to maintain, but would shuttle back and forth between flybys of Enceladus and Titan, repeating the cycle every few months.

The T9–T12 period of November 2005 to April 2006 marked a barren spell for the RADAR team, a data drought while the other teams continued to be deluged. RADAR was also uninvolved in the Enceladus adventures. (One reason RADAR has yielded so many key Titan results is that its team is focused almost exclusively on that one object.) But the opportunity to have some uninterrupted time to document the findings so far—to finally write up some papers on what was in the images— was not altogether unwelcome. For example, interpretation of the passive microwave emission from Titan's surface was making progress. The western boundary of Xanadu had been observed on TA, and in a different polarization on T8. This polarization information confirmed that Xanadu was not radar-bright because it was made of denser material, but rather because of some sort of textural difference. Meanwhile, other instruments continued to study Titan.

The UVIS experiment had a component that was to be used exactly once in the whole mission, on T9. This HDAC (hydrogen deuterium abundance cell) would measure the abundance of deuterium, the heavy isotope of hydrogen, in Titan's upper atmosphere. The deuterium abundance was difficult to measure any other way, even with the *Huygens* probe, but knowing it would help understand how much of Titan's atmosphere had been lost over time, and how. T9, on DOY (Day of Year) 360, the day after Christmas, was one of the highest "close" flybys of the mission, at some 10,000 km. The flyby included wide-area mapping by the optical remote sensing instruments, and the plasma instruments enjoyed a pass through the magnetospheric wake of Titan, much like that during the *Voyager* encounter.

T10, three weeks later, was also devoted primarily to optical sensing and plasma science. It included the first UVIS solar occultation of the tour. Since the Sun is a powerful UV source, much more so than a (more) distant star, it would allow more sensitive probing of Titan's upper atmosphere for methane, nitrogen, and other gases at high altitudes.

The T11 flyby at the end of February was set up as a gravity pass. The flyby was tuned to be low enough for Titan's gravity to bend the trajectory significantly but high enough above Titan so that the atmosphere wouldn't affect the spacecraft. The attitude thrusters were switched off, since they can lead to slightly unbalanced forces, which would contaminate the gravity measurement. In effect, *Cassini* was to behave as much like a cannonball as possible, its precisely controlled radio transmitters acting as beacons. And the bending of its trajectory would be measured with exquisite precision by radio tracking, to determine how mass is distributed within Titan. Does it have an iron core, or just a mixed iron/silicate interior?

This first gravity flyby may not tell us the whole answer, but a total of four such flybys are planned. Together they represent an ambitious experiment. There are two pairs of flybys, with one over Titan's pole and one over its equator in each pair. One pair is near Titan's periapsis and the other near apoapsis. The idea is that it may be possible to measure the tidal distortion of Titan's crust due to Saturn's gravity—an effect that is stronger near periapsis—and thereby infer whether it has an internal "ocean" of liquid water, or perhaps ammonia-water.

Some modeling work, by a French postdoctoral researcher, Gabriel Tobie, together with Christophe Sotin at the University of Nantes in France, and Jonathan Lunine in Arizona, showed that this internal ocean should persist to the present day (as other studies had also suggested) but, more intriguingly, that for much of Titan's history, the crust might have been thinner, an almost Europa-like 10 km or so. Only in the last half or one billion years had the crust thickened to closer to 50 km. In this thicker-crust stage, convection in the ice might slowly bring parcels of methane clathrate close to the surface, resulting in volcanic eruptions. If this history were true, it would have some interesting implications, not least that methane might not have always been present in the atmosphere. At that time, a billion or two years in the past, Titan might have been cooler (lacking the greenhouse effect due to methane), but its sky would not have been so hazy.

The T12 encounter went with little fanfare. This too was a flyby devoted to radio science, but radio science of a different kind. On this occasion, *Cassini* went directly behind Titan as seen from the Earth, and its radio antenna was aimed at the image of the Earth, refracted through Titan's atmosphere. (The radio beam is bent by about 2°.) This radio occultation was a repeat of the experiment made by *Voyager 1* all those years ago, the experiment that told us—correctly—what the density profile of Titan's lower atmosphere was. Without that information, design of the *Huygens* probe would have been much more challenging. However, a significant difference this time was that *Cassini*'s radio science system was much more sophisticated than *Voyager*'s, and could transmit simultaneously on three different frequencies—S, X, and Ka bands, with wavelengths of 8, 4, and 1 cm, respectively.

Another aspect of the experiment was "bistatic scattering." In essence, this is doing radar with a transmitter in one place and a receiver in another. The combination of different wavelengths might also provide some clues as to the scales of surface or subsurface roughness. Surprisingly, though, the different wavelengths were affected somewhat differently by their passage through Titan's atmosphere. The Ka band signal could not penetrate all the way through the atmosphere. The S and X bands passed through with little absorption. It would be important to understand why, not least in preparation for future missions.

Could it be that some gas was absorbing the short-wavelength radio waves—perhaps some of the nitrogen-bearing compounds like HCN? Another possibility was that cloud droplets were scattering the signal. The greater sensitivity of the short wavelengths to weather also affects the Ka band on Earth, and one has to accept that there is a higher risk of interruption by a rainstorm than when using longer wavelengths.

## T13: UNLUCKY FOR XANADU

The April 30, 2006, flyby, T13, was widely anticipated by the Titan surface community because it featured a radar study of the large, bright, leading-edge feature Xanadu, which had been Titan's most prominent and familiar landmark since it was first outlined by the Hubble Space Telescope in 1994. T13 would reveal why Xanadu was optically and radar- bright. *Cassini* had been performing reliably, and space scientists and engineers are not, by and large, superstitious. But this time, something went wrong. When *Cassini* was supposed to transmit its data on Monday morning, the Deep Space Network heard nothing. A spacecraft failing to report in is never good news. Could it be just some sort of easily fixed timing error, corrected after a controller slaps his or her head after working it out, or could it be a catastrophe like when *Mars Observer* or *Beagle 2* was simply never heard from again?

Controllers quickly switched the communications with *Cassini* to a two-way mode, where the frequency of *Cassini*'s transmission is locked to one sent up from the ground, and instantly the spacecraft responded— a huge relief! *Cassini* was no longer AWOL. What had happened, it seems, is that the ultra-stable oscillator (USO) that drives the one-way downlink had been switched off, but not by an erroneous command. Instead it seems its power switch had been flipped off by a random cosmic ray. When a charged particle, perhaps from a distant supernova, lances through electronics, it can cause a bit in a memory chip to flip from one to zero or vice versa. This happens perhaps a few times a year on *Cassini*. In a proportion of these cases, the corresponding unit is switched on, and so it might matter. But this seems to have been uncannily bad timing— for the USO to be tripped just before the downlink from a Titan flyby.

The spacecraft, then, was safe. Power-cycling (turning off and then back on—just as you might do with a crashed PC) restored the USO back to life. But what about the precious Titan data? *Cassini* is a powerful asset, and its time around Saturn is limited. Once its data recorders have been read out to the ground, there are new observations to be taken, which begin to fill the recorders again. In fact, particles-and-fields data often continue to be taken and recorded even while other parts of the recorder are being read out to the ground. And so, unless instructed otherwise, *Cassini* would begin to write over the old data it thought it had transmitted. *Cassini*, intelligent as it is, is not equipped to make value judgments on one science observation versus another, nor can it know if bad weather was degrading the signal in Madrid, or (as once happened just prior to a *Voyager* encounter) that a bulldozer had plowed through some cables at the Canberra ground station.

So if a downlink is lost, some special intervention is required to recover the data. On many Titan flybys, *Cassini* is sweeping inward toward Saturn, its rings, and its moons. Within a day or so there are new, genuinely unique opportunities to observe the rings with special geometry, or to see the jets of Enceladus. In such cases, there just isn't time to recover.

It was an agonizing wait, and the fallback window for downlink used only one of the Deep Space Network's smaller antennas, so the data came back at only 34 kbit/s—half the speed of a phone modem, or one-thousandth of the speed of a typical modern broadband Internet connection. Without the recovery plan, 70 percent of the data from the flyby would have been lost. Thankfully, with it in place and activated promptly, only 17 percent of the data was lost. The experience with T7 had been a good lesson.

It was a sobering reminder, not only of how hard *Cassini* was being pushed and how unforgiving a space mission can be, but also of how much behind-the-scenes work goes on. Dozens of backup plans have to be worked out and ready to go, but are thankfully never used. Even though *Cassini* is working at Saturn for four or six years, some days in that period are precious indeed.

## XANADU REVEALED

Sure enough, as the earlier indications had suggested, Xanadu was rough. When the radar image came rolling out, it revealed that most of the 2,500 km extent of Xanadu was rugged, full of broken or gullied terrain. Nothing looked desperately volcanic, nor was tectonics strongly in evidence, with only a few straight edges to features. The landscape, cut and broken, looked much like radar images of the Himalayas, although it was still unclear what made it so rugged.

A couple of dramatic, long, and branched river channels wound their way across the width of the image—implying channels some 200 km or more in length. There were also a couple of features that resembled impact craters. One in particular was a puzzle. It was the same size as the

Figure 6.17.   The rumpled hilly structure of Xanadu is evident in this portion of the T13 radar swath, which shows about 200 km top to bottom, with radar illumination from the top. It seems that dark material forms flat areas between the hills, which have some straight-edged patterns suggestive of tectonics. Preliminary estimates of the height of the hills are up to 1 km. (NASA/JPL)

crater Sinlap identified earlier, but instead of being a neat, symmetric pit with a flat floor, this new feature had a central peak. Evidently, when this crater formed, there were rather different structural characteristics— Xanadu was clearly a different kind of place on Titan.

Independent of *Cassini*, nearly a hundred scientists were gathered in Pasadena for a meeting of the Outer Planets Assessment Group (OPAG), a forum for organizing consensus on the missions and approaches to be advocated for future exploration. A thirty-foot-long printout of the T13 swath was rushed down to the meeting site for the attendees to gape and point at during their coffee breaks.

Although most of Xanadu had not been seen optically at highest resolution, its western boundary had been seen quite well. As with the landing site location, there were correlations between the ISS and radar images, but also some striking differences.

Some cases in point were several bright regions in the Shangri-La dark terrain, to the west of Xanadu. One of these was a ring of bright patches, named Guabanito. This was suspected to be an impact structure, perhaps buried under dunes, much like the Arounga impact crater in Chad. It was hoped that its nature would be more obvious in the radar image, that perhaps much of the dark floor would prove to be just a thin veneer of sand or sludge, and that more of the geological structure would be revealed by the radar. There was to be no such luck. It looked just the

Figure 6.18.  Rivers to a Sunless sea? The narrow, winding bright channels seen in this segment of the T13 SAR swath at the western margin of Xanadu are the longest river channels identified so far on Titan. (NASA/JPL)

same in radar. A little to the north, however, the complementarity of the two approaches was shown in the bright splotch named Kerguelen Facula. This was dissected by a St. Andrew's cross of dark material. The radar image showed one of these dark lanes with the characteristic topographic shading that indicated it was a valley. Dark stuff filled the valley.

One of the most appealing features was that known officially as Shikoku Facula, although informally better known as Great Britain. This too was crossed by some dark, sinuous tracks, and surrounded by the dark expanse of Shangri-La. To no one's surprise, the dark plains could be seen in the radar data to be more "cat scratches" or dunes, but not everything in the image was anticipated. A 60 km round, almost polygonal, dark spot at the north of Shikoku (roughly at Wick, if you use a map of Great Britain to work out where we mean!) stood out in the radar image, but was virtually invisible in the optical map.

Larry Soderblom made a color composite image, using low-resolution VIMS color to shade the higher-resolution brightness map. This technique showed some correlations very well—correlations that had been hinted at elsewhere. By making the colors correspond to the different windows (blue for 1.3 microns, green for 1.6 microns, and red for 2 microns), they would indicate compositional (or possibly particle-size) variations across the scene, whereas the radar image indicated the landscape itself.

The eastern edge seemed to have a blue tint, much like the *Huygens* landing site, and the dunes tended to be a sort of brown. The dark spot had a greenish tint. These sorts of correlations would start the matching

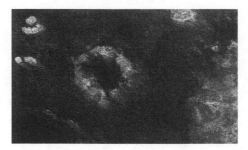

Figure 6.19. The western edge of Xanadu is visible at the right of this radar image. The circular feature (80 km in diameter) is named Guabanito, and looks very similar optically (see figure 4.06). The alignment of the dark streaks (presumably shallow dunes) is quite different from those in Belet to the west of here. Perhaps the winds are diverted by Xanadu. (NASA/JPL)

of composition to process on Titan, and allow generalization of terrain types—for example, judging areas only seen with VIMS to be dune fields, even if the dunes had not been observed by radar, and vice versa. Unfortunately, the best VIMS/RADAR overlap should have been on the eastern side of Xanadu, but the radar imagery had been lost in the downlink problem. However, the Shikoku results were promising.

## T13: CRITICAL MASS

The data obtained on T13 by no means meant we understood Titan. But it did mark the onset of what was likely to be an ever more significant feature of the mission: the synergy between different instruments. And similarly, in the first part of the *Cassini* mission, the modest coverage by radar and optical techniques gave only partial views of isolated areas on Titan's complex surface. But, as the coverage by both ISS/VIMS and RADAR steadily increases, the number of areas studied by both will increase quadratically.

Furthermore, as the broader scientific community comes to grips with the *Cassini* data filling up the archives of the Planetary Data System, new and different approaches will be applied to it—new correlations, new ways of extracting information on texture or topography and composition. And new models, too, will put the data into a broader context, explaining why the river valleys are where they are, and what different weather may explain the different amounts of river incision in different places.

Figure 6.20. A close-up of the T13 radar image of Shikoku, which looks less like Great Britain than its optical counterpart. Dark streaks are sand dunes, except perhaps the dendritic dark features forming a forked network about two-thirds of the way up. The dark "circle" at the top is much more prominent in the radar image than optically, illustrating the usefulness of having both data types available. (NASA/JPL)

Titan is the first world to be mapped optically and with radar, and studied in situ with a probe, at the same time. In some ways, this leapfrogs the usual progression of flyby—orbiter—probe that has characterized the exploration of the terrestrial planets.

Venus's gross topography and distribution of radar reflectivity (comparable with what can be achieved with the *Cassini* radar scatterometer) were obtained by ground-based radar and by the *Pioneer Venus* orbiter in the late 1970s and early 1980s. This was also about the same time that the first pictures were returned from its surface by *Venera 9* in 1975. But a global imaging radar survey had to wait until the late 1980s with *Venera 15* and *Magellan* in 1990. And only in recent years has it been realized that information on the surface can be gained in the near-infrared, in windows between $CO_2$ and $H_2O$ absorption bands, just as $CH_4$ windows on Titan can be used to sense the surface.

Mars's surface, of course, is visible from Earth, and had been mapped optically by *Mariner 9* and *Viking* in the 1970s. But the topography data that allows its fluvial and glaciological processes to be understood became available only with the Mars Orbiter Laser Altimeter (MOLA) dataset from *Mars Global Surveyor*, a quarter of a century later.

*Cassini*, sadly, is not a Titan orbiter, and so, even in its extended mission, it will not get anywhere near a global map. But in one fell swoop, if that term can be applied to the four-year-plus campaign of scientific attack on Titan, Titan has been transformed from an object of speculation to a planetary world with its own set of processes and observable effects.

# 7. Where We Are and Where We Are Going

As we put the finishing touches to this final chapter, the *Cassini* mission continues. But we have to draw a line somewhere and take stock even though new findings may be just around the corner—findings as thrilling and intriguing as those from the last Titan flyby we can include, T16 on July 22, 2006.

## T16: LAKES AT LAST?

Titan had seemed almost defiant in concealing evidence of present-day surface liquids. The Ontario feature seen optically near the south pole was compelling but not completely persuasive, and the RADAR look at the south pole on T7 had been thwarted. The models that suggested Titan's high latitudes should be damp were at least consistent with it being somewhat dry—and largely covered in sand dunes—at low latitudes. There was even a certain logic, given the rates of evaporation in summer, to lakes being dominant at the winter pole. And so for many, T16, the first of numerous planned RADAR encounters near Titan's north pole (*Cassini*'s trajectory does not, in fact, cover the south polar regions often), was the "last chance" for lakes of ethane and methane. If they weren't seen in Titan's arctic, perhaps they wouldn't be seen at all.

T16 did not disappoint. The RADAR swath, acquired on July 22, 2006, showed a dramatic collection of what had to be lakes at the highest latitudes covered (about 80°). These were shaped like lakes (as was the south polar feature), but there were many more of them. The superior resolution of the radar images showed several to have channels draining into them, although interestingly, most did not. This was consistent with Pascal Rannou's climate model, in which a steady ethane-rich drizzle, rather than sudden methane thunderstorms, may characterize the high-latitude climate, so that the intense erosion needed to form channels does not often occur. Several of the lakes seemed to have "bathtub rings"— abrupt changes in slope that paralleled the edges of the lake, suggesting that the lake had once been filled higher. None of these factors absolutely meant that they were filled with liquid today—they could be lake beds.

But they were pitch black to the radar—much darker even than the dark spots and Si-Si seen on TA. They were virtually indistinguishable from nothingness, as if these areas had swallowed the radar energy completely. In principle, there were three possible explanations for surfaces behaving this way, although all required some geological process to form the lake-shaped feature in the first place.

First, and least likely, they could be made of something dense, such as ice, sludge, or even metal, that was very smooth. Metal didn't seem likely, and for ice or sludge to behave this way, it would have to be devoid of internal imperfections that would otherwise give some diffuse reflections. That seemed rather improbable, and could also be dismissed another way. If they were mirrors like this, they would appear like cold space to the radar's passive radiometer mode. They didn't; they were strongly emissive.

Second, they could be filled with some low-density absorber like soot, much like the "Stealth" radar-absorbing region on Mars, which is believed to be a thick deposit of fluffy volcanic ash. But this would mean something made the lake-shaped pits, and these were then filled with some fluffy material, which somehow did not get blown around into dunes but stayed perfectly flat.

Or third, by the rule of Occam's razor, they looked like liquid hydrocarbons filling lakes because that's what they were.

This blackness required that they were very smooth, that there were no large waves on the surface. Could Titan's winds of 0.5 m/s, enough to make sand dunes, be enough to kick up waves on such lake surfaces? Some work that had been done on the topic before *Cassini*'s early results suggested that it might be a waste of time, but the T16 observation gave the question new prominence.

A feature of the Earth sciences is that there are many complex processes that both have important effects on us and are close enough to

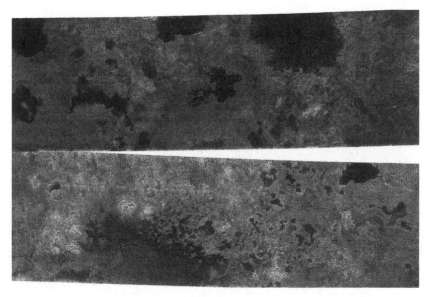

Figure 7.01.   Lakes at last. This radar view from the T16 north-polar flyby shows what appears to be a large collection of lakes, varying in size, brightness, and (perhaps) depth of liquid. (NASA/JPL)

observe in detail. The generation of ocean waves by the wind is a prime example. From a purely physical and mathematical point of view, the problem is hideous. The drag exerted by a wind on perfectly flat water will cause the surface to become unstable and little ripples will appear. It is the surface tension of water—the "skin" that stops pond-skating insects from getting wet—that pulls these waves back down. But the rippled surface is rougher, and so drags more momentum down from the airflow above and the waves grow. There is thus a complex feedback between the airflow and the water surface, and things get worse as different waves interact: the waves become too steep and break, and so on. The generation of water waves is a problem with which nineteenth- and early twentieth-century mathematicians such as William (Lord) Kelvin, Harold Jeffreys, and Horace Lamb all grappled, with only modest success even for the simplest of cases.

So scientists throw up their hands and take the expedient, and perfectly effective, approach of empiricism. It doesn't matter how the waves grow and interact theoretically—we just put instruments on a lot of ocean buoys. We acquire huge numbers of datapoints of "windspeed $x$, wave height $y$" and choose a mathematical curve that goes through the middle of the cloud of points on the graph. Then, given a predicted wind speed, one just applies the equation of the curve and gets a decent estimate of what the wave height will be. (Actually, of course, it is a bit more sophisticated than that: the wave height depends on the width of the stretch of ocean—the "fetch"—over which the wind has had to act, and the wavelength will change with time too. Various empirical relationships exist for all these.)

The problem is that such empirical relationships are essentially worthless in an entirely different environment. Some half-hearted efforts were made to translate the empirical relations for Earth to Titan, by scaling the wave heights by a factor to take into account the different gravity. But that does only half the job. What about the different density of the air, the different density, viscosity, and surface tension of the liquid? And since the whole wave-growth process has many feedbacks, it is very non-linear, so it wasn't even obvious that the gravity factor used was the right one. What was needed was an experiment.

..........................................................................................................

## RALPH'S LOG, AUGUST 2003

### The Mars Wind Tunnel, NASA Ames Research Center

This is just the kind of experiment I like doing. Something messy—the sort of garage enterprise that no one would want to taint an actual lab with. I like something that sounds simple—even silly—but that will tell us something genuinely new and is small enough in scale that I can just go ahead and do it, and not have to write a big proposal to get special equipment.

Visiting the University of California, Santa Cruz (UCSC), I had encouraged a grad student, Erin Kraal, to think about seas and shoreline erosion on Titan. She was, in fact, working on Martian shorelines at the time, with her UCSC advisor Erik Asphaug and Jeff Moore at NASA's nearby Ames Research Center. She worked on making a mathematical model of how waves of a given height will erode a shoreline to produce cliffs, but the question remained, How big would the waves be? This was as impossible a question for ancient Mars as it was for present-day Titan; some decent guesses could be made at what the wind speeds might be, but how big would that make the waves? Working with Erin kept me thinking about the waves, and opened the way to doing an experiment.

The key was MARSWIT—the Mars wind tunnel, operated by Arizona State University at NASA Ames. I had heard of this facility before—it was used to simulate the formation of wind ripples and sand dunes on Mars. I was familiar with wind tunnels from my undergraduate degree in aerospace systems engineering, and I assumed that MARSWIT was a closed-circuit tunnel that could be pumped down to a low pressure.

Officially, the tunnel is for research funded by the NASA Planetary Geology and Geophysics program. Other users are supposed to pay a fee. But we negotiated that since Erin was doing NASA-funded research, we could perform at least a trial experiment for a couple of days on a no-guarantee basis. We make a scouting visit to MARSWIT—I always find plenty of interesting people to talk with at Ames anyway.

The facility is not what I expected. Rather than a closed loop, the tunnel is open—just a long duct with a fan at one end. And this 20-m-long tunnel sits in a big concrete building (an old test chamber for rocket upper stages); all the air is pumped out of a huge section of the building! Massive steel doors isolate the vacuum chamber from the outside, and the corridors inside. An inch-thick glass window allows a view into the chamber from the control room, and slats would flick into place to limit the destruction if the window failed. The operators, nerds to a man, had taped a picture of Arnold Schwartzenegger's character in *Total Recall*, clinging to a pipe as the air is sucked out of his Mars base—just how the control room would look. Who knew Arnold would become the governor of the state in a couple of years!

We set up for a week of experiments, and I arrive at San Jose with some very odd-looking luggage. Not least was a big plastic tray. I have a bag full of electronic parts—ultrasound and infrared sensors to measure the surface height of the liquid.

At first, we just try the sensors out and run the tunnel at room pressure. It makes little waves, which we can see in the data. We tweak the sensors a little (measuring things in the real world, I find nothing ever quite works as expected) and hope we'll be able to run at low pressure.

The awkward aspect of being a "piggyback" customer is that one is at the mercy of the steam plant's schedule. It takes an enormous amount of power to pump all the air out of a building, and this comes from a huge steam plant across the street (whose main function is to pump the air out of the arcjet plasma tunnels used to test heat shields elsewhere at Ames). And after a hopeful couple of hours, it emerges that the steam plant is not going to run that day.

Rather than waste the afternoon, we make a dash to a nearby hardware store and buy four gallons of kerosene. Running the wind tunnel with the doors to the building open to get rid of the smell, we find that, indeed, waves in kerosene are larger for the same wind speed than they are in water. That, at least, is a result.

The following day, we get some low-pressure results. I return to Tucson, leaving Erin to run the experiments for a couple of days. She bakes a motivational apple pie for the guys in the steam plant—a master stroke. Somehow the steam plant becomes very responsive to our needs, and in my absence, Erin generates a great set of data, with runs at different speeds at various pressures.

When I return the following week, on the way to the DPS meeting, we do another few runs and pack up. We try a very low-pressure run—10 mbar or so, about the highest pressures that exist on Mars today. The thin air fails to make any perceptible ripples, and the cooling by evaporation, the water being close to boiling at these low pressures, is enough to chill it to the point that a sheet of ice forms over the tray. Well, that's that, then. Waves are pretty hard to make on Mars.

As we analyze the data some weeks later, to turn the stream of numbers from the sensors into a quantitative relationship between wind speed, pressure, and wave height, it becomes clear that there is a very strong sensitivity of wave height to

Figure 7.02.   A "kitchen sink" experiment in a unique facility. The Mars wind tunnel (MARS-WIT) is housed in a huge airtight building at NASA Ames Research Center (*left*). It contains a small open-circuit wind tunnel (*center*), in which an instrumented tray of water and hydrocarbons (*right*) was installed to study wind-wave generation in extraterrestrial environments.

pressure, but not the linear, well-behaved proportionality one might expect: waves basically are completely suppressed below 600 mbar, where the air density is only 60 percent of that at sea level on Earth. Titan, on the other hand, at 1,600 mbar and four times the air density, might therefore be much more effective at generating waves. And that's in addition to the weak gravity. It's a neat result, and a fun experiment that my colleagues appreciate.

Probably someone, somewhere, doesn't appreciate these experiments. A little kerosene spillage occurred in the rental car, which continued to smell strongly thereof. Let me hereby record my apologies to the next person who rented that car at San Jose airport.

## THE END OF THE BEGINNING

It is hard to break off from the ongoing story of discovery at Titan, but break off we must. As this book goes to press, *Cassini* scientists anticipate another busy couple of years of the nominal mission, and a mission extension of at least a couple of years after that. All of this could change. Fate may catch up with *Cassini* and cause some instrument or crucial system to fail. Even being a productive asset in space is not always enough to avoid budget cuts. Though *Cassini*'s science operations were shaved to the bone years before (by some reckonings, the science teams are 50 percent underfunded), *Cassini* is a big item, a tempting target for budget planners who may not appreciate the damage that even what looks like a small, incremental cut can cause to a thinly stretched team.

What we may have to look forward to in the current plan includes the following highlights. With T11, T22, T33, and T38, the radio science team should have a set of data that nails down the internal structure of Titan, setting the context for any volcanic processes and perhaps indicating how thick the ice crust is.

Atmospheric results will accumulate, building up a multidimensional picture of how the haze, gases, temperatures, and wind vary with latitude, altitude, and time (and, for that matter, perhaps longitude too).

Most of Titan's surface will be mapped somehow—the sub- and anti-Saturn areas seen so often during *Cassini* flybys, and low latitudes in general, should be quite well covered in the near-infrared and by radar. The southern hemisphere as a whole, in autumn sunlight during the nominal mission, should be fairly well imaged by VIMS and ISS, while the northern hemisphere is bracketed by many radar swaths. With all this in hand, we should have an understanding of the gross distribution of geological features and be able to say, for instance, whether there are any dunes at high latitudes, lakes in summer, volcanoes or craters in particularly old or young provinces. It seems certain there will be a long list of places we want to see again more closely.

By the end of the nominal mission, the RADAR instrument will have mapped about a quarter of Titan's surface with its high-resolution SAR mode. This should be enough, except in some poorly covered areas such as the southern part of the trailing hemisphere, to get a good inventory of the various terrain types, and to understand what the global patterns of fluvial, aeolian, and volcanic activity might be.

## ALL CHANGE

An extension to *Cassini*'s mission at Titan will be particularly interesting around 2010, the northern spring equinox (in fact, one Titan year after the *Voyager 1* encounter in 1980). The reason is that, as the Sun moves north and crosses the equator, the distribution of sunlight on Titan, which drives the Hadley circulation), will reverse. Instead of the high-altitude winds going from south to north, and dragging haze with it, for a couple of Earth years, there may be an unusual (for Titan) Earth-like symmetric circulation pattern with warm air rising at the equator and descending in both hemispheres. Then the more usual "pole to pole" Hadley circulation will resume, this time going from north to south.

Another phenomenon that occurs at the equinox season is shadows. Titan can pass into eclipse behind Saturn. This is almost like a laboratory experiment in which the effects of sunlight on Titan's upper atmosphere can be isolated. There are so many correlated combinations of orientation of plasma flow and sunlight on Titan that separating the two effects is difficult, but as Titan goes into eclipse, there is a sudden change from Sun to no Sun, while the plasma environment remains the same.

It will also be an important time to study the rings, when they are edge-on to the Sun and subtle variations in thickness, and bending waves or other warping will become especially visible. The grazing sunlight on the rings is also important for seeing the spokes that were first observed by *Voyager*, but have been almost invisible while the Sun has been high above the rings.

## NEW TELESCOPES

Although *Cassini* has the best vantage point for observing the Saturnian system—from within—the view from Earth will nonetheless improve. New developments in telescopes and instruments continue to sharpen the Earth-based view.

HST's successor, the James Webb Space Telescope (JWST), is set to launch sometime in the next decade. Its mirror, some 6.5 m in diameter, will be over twice the size of Hubble's, giving it better light-collecting ability and higher resolution in principle, although that may depend on it being given the rather specialized ability to track solar system targets. JWST will make a spectacular series of observations, including images of Titan, from 0.7 microns to 5 microns with its near-infrared camera (NIRCAM), spectra with resolution of 1,000–3,000 over the same wavelength regions with the near-infrared spectrometer, and both images and spectra in the range of 5–27 microns with the mid-infrared instrument (MIRI). The spectral capabilities of the near-infrared spectrometer are much superior to those of *Cassini*'s VIMS, and JWST's large mirror will give enough signal to search for the faint traces of organic compounds that have so far eluded identification. JWST, and the ever-growing arsenal of ground-based telescopes—with mirrors 20–30 m in diameter and images sharpened with near-infrared adaptive optics systems—will continue to monitor how Titan's cloud patterns change with time.

On the ground, the 10-m-class telescopes like Keck and Gemini may give way to arrays of telescopes, almost magically connected to give the resolving power of much larger telescopes, and possibly much, much larger telescopes. These telescopes are still in their early planning stages, but the OWL (Overwhelmingly Large Telescope), many tens of meters across, could yield improvements as significant as the jump from the 1980s 3 m class to the turn-of-the-millennium 10 m class.

At other wavelengths, too, new tools will probe Titan. The VLA radio telescopes will be improved, to form the Expanded Very Large Array, with improved sensitivity and resolution. And there will be improvements, too, at shorter wavelengths, in the millimeter-wave range, where observers in the 1990s profiled a number of compounds, such as HCN and CO, in Titan's atmosphere. A facility under development operating in this wavelength range is ALMA, the Atacama Large Millimeter Array (oxymoronic as that sounds). This observatory will feature a large number of shiny dishes, spread out over the high, dry Atacama Desert in Chile.

With the high resolution afforded by the array with these telescopes synthesized together—better than Hubble achieves at optical wavelengths—variations in temperatures and gas abundances with latitude and altitude may be monitored and the seasonal fluctuations tracked. All these Earth-based observations will help refine models of Titan's climate and winds, which will be useful for the next step: flying on Titan.

## THE EASIEST PLACE TO FLY

Serious contemplation of Titan exploration began in 1973, at a workshop at NASA Ames. Even then, before we knew any details at all, the idea of flying a balloon on Titan was discussed. After all, Titan had an atmosphere, so why not? The *Voyager* encounters confirmed that a balloon or airship would indeed be a quite viable concept. In Titan's thick atmosphere, a balloon could be made rather smaller for a given carrying capacity than it would need to be on Earth.

A case could also be made for heavier-than-air platforms to explore Titan. These would benefit not only from the thick atmosphere but also from the low gravity. A nuclear-powered airplane could explore Titan for years, although accessing the surface would be a challenge. In the last five years, however, great strides have been made in unmanned aerial vehicles (UAVs) on Earth, to the point where they can fly autonomously across the oceans (or, for that matter, fire missiles at people). A more technically ambitious idea than an airplane would be a helicopter. Although more complicated, a helicopter has the obvious advantage that it could land more easily on the surface, to do chemical sampling or even take the seismic pulse of Titan.

But lighter-than-air ideas retain an appeal: they are simple, and somewhat fail-safe in that they should stay floating if they happen to lose power. A favorite option in the early part of the decade was an airship, able to move around more or less at will on Titan, and able to sample surface material by means of some sort of tethered grabber. If launched in 2010 or soon thereafter, it might be directed to Titan's north pole, where, by the time it arrived in 2017 or so, the season would be midsummer. That meant that the pole would be not only in continuous sunlight but also in basically continuous view of the Earth. With a steerable dish antenna inside the airship envelope, or perhaps using an electronically steered phased-array antenna, it could communicate directly with Earth. Now, at Titan's great distance, even a steered antenna would be able to pipe data down at one or two kilobits per second, a quarter of the data rate from the *Huygens* probe to *Cassini*. And it could do that not for two and one-half hours, but in principle for twenty-four hours a day, week after week.

The beauty of communicating directly to Earth was that the mission would not need an orbiter. Instead of a series of orbital insertion maneuvers and separation events, an airship could just be thrown fast at Titan and enter directly.

## HOT AIR ON A COLD WORLD

An airship on Titan could use helium for lift, although there would be no reason why one couldn't use slightly lighter hydrogen instead, since on Titan, where there is no oxygen, the hydrogen couldn't catch fire. Some more recent ideas involve hot air balloons. It is easier to make these go up or down by modulating the temperature of gas inside the balloon.

Figure 7.03. A rosy future for Titan exploration? An artist's impression of an autonomous airship exploring a methane vent on Titan. Saturn is visible in the background, and a faint rainbow is cast in the methane steam plume. The disks on the airship envelope form a steerable array antenna to transmit data directly to radio telescopes on Earth. (Mark Robertson-Tessi and Ralph Lorenz)

On Earth this is done by using propane burners. On Titan the waste heat from a radioisotope generator could be used.

The thermoelectric devices that convert the heat from a radioisotope power source (RPS) into usable electricity are not very efficient—typically only 5 percent. And so, what is specified as a source of 100 W of electricity is actually putting out about 2 kW of heat. In fact, getting rid of the surplus heat is often a challenge for spacecraft designers, especially when the source is encased deep inside some entry heat shield. Extra fluid loops and radiators often need to be installed. Normally, this heat is just wasted, or perhaps a little is tapped off to help keep a spacecraft warm, but on a Titan balloon or Montgolfière (the name the Montgolfier brothers gave their balloon; both pronunciations seem to get used), the cooling fins of the RPS would be placed at the neck of the balloon, putting enough heat in to keep the air 15 K or more above the ambient temperature. Jack Jones of JPL showed that ballooning this way on Titan is many times more efficient (in terms of lift per watt of heat) than on Earth. Probably the skin of the balloon would need to be double-walled (as some long-duration balloons on Earth are) in order to keep the heat in enough to inflate the balloon as it descended by parachute when released from its entry shield.

A large vent valve would also be placed at the top of the balloon, to allow the warm gas to be dumped to cause a descent. This would allow the balloon to get close to the surface for more detailed inspection of areas of interest, or even to sample surface material with some sort of harpoon.

Julian Nott, a British balloonist who holds seventy-six world records and demonstrated how the ancient Peruvians might have used hot air balloons to direct and observe the Nazca lines—great drawings on the landscape, visible only from above—has embraced the Titan ballooning challenge. At a probe workshop in Pasadena in June 2006, he suggested that one might use a sort of small rocket engine to quickly fill a balloon with hot air as it descended from its entry shell, giving it instant buoyancy, rather than landing on the ground first or waiting for the slow heat of a radioisotope source to warm it up. Whether this will turn out to be an efficient solution will require some detailed study, but that is not the point—ballooning on Titan is capturing the technical imagination of lots of smart people. And Titan as an environment is comfortable enough that there are lots of technical possibilities. Engineers aren't "walled in" to a single way of doing things.

What might a balloon measure? It would take pictures, of course. It could drift for a year or more, taking sharper pictures than *Huygens*, and thousands of times more of them, giving us an "airplane window" view of Titan's varied landscape. A ground-penetrating radar is another idea—one that is able to measure the depth of the lakes, and perhaps the depth of the sands in Belet and Shangri-La, and possibly revealing the structure of buried craters like Guabanito.

And of course, measuring the composition of the organic materials on the surface is a key objective. Just how complex do molecules get when Titan's tholin is exposed to liquid water for thousands of years?

One can't land at and sample every spot on the surface, so some way of classifying the chemistry remotely is needed. Near-infrared spectroscopy may offer some clues—*Cassini* VIMS data are already hinting at some compositional variations. But a rather interesting idea came from work by Rob Hodyss at Caltech in 2003. In studying tholins made in the laboratory, and the compounds formed by the interaction of tholins with water, he discovered that tholins fluoresce. That is, if illuminated by ultraviolet light, they glow. You can see the same effect in a nightclub. Fluorescent materials are added to laundry detergents to give that "whiter than white" appearance, and so white shirts glow impressively under UV lights. The organic compound and antimalarial agent quinine is also fluorescent, so gin and tonics glow a pale green under UV illumination. Interestingly, tholins that have been mixed with water and then frozen into the ice glow a slightly different color from pure, unwashed tholins. And so, a balloon near Titan's surface equipped with a UV searchlight (remember, all the Sun's UV has been absorbed by methane and the haze at high altitudes on Titan) could spot tholins exposed to hydrolysis, and hence the most likely sites for the most interesting prebiotic chemicals, by the way they glow under UV illumination.

The scientific community has much work to do to work out the best measurements to characterize Titan's surface materials. On Mars, it is easy. Almost any organic molecule would be interesting, since the Red Planet's oxidizing soils break organics down quickly. Titan is awash with organics, and so ways of measuring the most interesting ones are needed.

Most thinking has, not unnaturally, been directed toward what Titan's chemistry can tell us about how chemical systems can become progressively more complex and ultimately lead to life. We don't know how far this process went on Titan. Although nothing is impossible, it seems unlikely that life has evolved on Titan.

It has been argued that Titan's interior (the liquid water layer) is not in principle uninhabitable, since there could be abundant dissolved nutrients there. So, what if life came from elsewhere to Titan? This question was addressed by dynamicist Brett Gladman of the University of British Columbia in 2006. He noted that rocks ejected from Earth during impacts could make their way to Titan, much like the Martian meteorite ALH84001 that was found in Antarctica came to Earth. And Titan's thick atmosphere would give them a soft landing. Although the odds are somewhat astronomical, microbes from Earth could have been delivered to Titan's surface. If they happened to land during Titan's early years, when the water—ammonia ocean was exposed to the atmosphere, perhaps they might have flourished. You can never say never.

......................................................................................

## RALPH'S LOG, 2001

Tucson

Radiocarbon on Titan. It is comparatively rare in modern specialized science, and is often the mark of good science, that an idea can be completely encapsulated in three words. Tim Swindle, a meteoriticist at the University of Arizona, began considering techniques for determining the geological ages of materials on Mars, Europa, and Titan. When he suggested that cosmic rays impinging on Titan's nitrogen-rich atmosphere would create carbon-14, as they do on Earth, the idea clicked in my head. Of course!

Tim and I, together with Tim Jull, a world expert in radiocarbon measurements using the sensitive technique of accelerator mass spectrometry at the University of Arizona (and, like myself, a Brit by birth—we seem to get everywhere), and Jonathan Lunine, began to work out the details. In the process, I began to wonder how much of a lethal dose I have accumulated crossing the pond while working on *Cassini*.

The essence of the situation is this. Energetic cosmic rays, flung out from distant supernovae, for example, can do a number of things when they strike matter. One effect is to tear apart molecules, creating ions or free radicals. These radicals are what can attack DNA in living things, causing mutations. Another effect is to dump charge into semiconductors, causing "snow" in digital images or random errors in computer memories. But sometimes, just sometimes, the miracle of nuclear alchemy occurs. Cosmic rays hit a nucleus and transmute one element into another.

An awful lot of the nuclei that a cosmic ray lancing into the Earth from space will hit are those of nitrogen. Nuclei of nitrogen struck in this way tend to turn into carbon-14. In chemical terms, this acts much like any other carbon on Earth, and gets assimilated as carbon dioxide into the tissue of trees and other living things. But carbon-14 is radioactive, decaying with the emission of a beta particle (of a characteristic energy that can be detected by a fairly simple detector if you have enough carbon) with such a probability that half of them do so within about six thousand years. This decay means the fraction of carbon that is radiocarbon in a sample of once-living material like wood decreases predictably with time and makes it possible to do radiocarbon dating.

Cosmic rays produce showers of secondary particles, notably muons. These are the dominant contributor to the background radiation measured at most points on the Earth's surface. This particle flux increases as one goes higher in the atmosphere. In Tucson, at an altitude of 700 m, a typical count rate for a small Geiger counter is a few counts per minute. Close to sea level, it is typically a factor of two or three smaller. In a commercial airliner on a transatlantic flight, at 10 km altitude, the count is twenty times higher. In ten hours the radiation dose is comparable with that received in a chest X-ray.

There is an altitude range over which the cosmic ray collisions tend to occur. At the very highest altitudes, there are comparatively few atoms in every cubic meter of air for the cosmic rays to hit, and so the rate is low. At the deepest levels in the atmosphere, the mass of gas above has already absorbed most of the cosmic rays. And so somewhere in between, there is a peak production rate. This occurs on Earth at altitudes of a few kilometers. On Titan, with its distended, more massive atmosphere, the production peaks around 70 km, well above the surface. It had been calculated that this cosmic ray absorption peak might produce a detectable increase in the electrical conductivity of the atmosphere around this altitude, as the electrons torn off by cosmic rays float around before recombining with an ion or molecule.

The Arizona team calculated that the production rate of radiocarbon should be a little (a factor of four or so) higher than the terrestrial rate, which is about two atoms per square centimeter per second. Because an atom of radiocarbon will decay with a half-life of about six thousand years, it follows that the quantity of radiocarbon would build up until there are so many that they decay as quickly as they are formed. This happens around five hundred billion atoms per square centimeter.

To balance this production of several per second per square centimeter, there must correspondingly be several decays per square centimeter of surface. But this can mean a lot of things—several decays spread through a kilometer-deep layer of liquid and sludge hydrocarbons on the surface, or dispersed through the tens of kilometers of the troposphere. In these cases, the handful of disintegrations per second would be hard to find and count.

But what if the radiocarbon, being formed in the lower part of the haze layer, tended to attach itself to the haze particles? Even though the haze looks thick, in reality it corresponds to a layer only a few microns thick. If most of the radiocarbon were to be incorporated into the haze, then "fresh" haze, falling from the atmosphere in a thousand years or so—well under the half-life of radiocarbon—would have a fairly high proportion of radiocarbon in it, perhaps one part per billion. This would be enough to make the material itself quite radioactive.

Although this stuff wouldn't be radioactive to the extent that refined radioisotopes are, it would be hundreds of times more radioactive than the most radioactive organic material on Earth (Brazil nuts; the nut trees take up a lot of calcium from the soil, and in so doing also incorporate a fair amount of radium, such that Brazil nuts give off about one decay per second per gram). It would be something easy to detect with a simple radiation counter.

Perhaps some day in the future, a robot airship traversing Titan's landscape may use a radiation counter to measure whether some deposit of organic gunk in the banks of a streambed is some ancient seam cut into by the stream, or was left behind when a methane storm just a thousand years ago washed material down from a wide area and concentrated it here for our convenience.

..................................................................................................

## SURFING THE TIDE

Simulations by Tetsuya Tokano and Ralph Lorenz in 2005 showed how a balloon on Titan would wander over a wide region and not just stick to a line of latitude, as was thought in the late 1990s. In fact, just how much the balloon would drift north or south depends in a rather interesting way on where it is. The tides cause a pressure bulge to sweep around Titan, causing winds with a substantial north—south component that act at a point on Titan for a couple of Earth days, a fraction of a Titan orbital period, before the tidal bulge sweeps past. How far the balloon is displaced depends on how long it spends in the tidal bulge. At the lowest altitudes where the zonal wind is weakest, the balloon almost sits in place and the bulge sweeps over it briefly, giving a small displacement. Conversely, at high altitude, the fast zonal winds carry the balloon through the bulge quickly, again yielding a small displacement. But at some intermediate altitude, the zonal wind carries the balloon along at just the same speed as the tidal bulge, and so the balloon stays in it, drifting north or south all the time. It's just like surfing—match your speed with the wave and it will sustain a large motion, whereas a buoy bobbing in one place or a speedboat lancing through a wave will barely feel any effects.

One interesting feature of the wind field computed in the models is that, in some seasons, the zonal (west—east) flow is reversed at low altitude—as indeed was observed by the *Huygens* probe. This means that a balloon that can control its altitude can first reconnoiter a site by drifting overhead and taking pictures, which are sent back to Earth. At the right altitude (a few kilometers), the balloon does not drift too far during the time taken to transmit them, plus a few hours for analysis on the ground. If the area looks to be of special appeal to scientists, the balloon could be commanded to descend to low altitude, where it would backtrack and sample a designated site.

Of course, many details need to be worked out—for example, how best to mesh the limited onboard intelligence of the balloon (which can respond immediately) with the more formidable intellect on the ground, which is two hours light time away. And at present, the tidal wind model is just a theory. But as the models are progressively tuned to include observations of clouds on Titan, and the orientation of the sand dunes, their predictive capability will improve.

## WHERE TO?

Whether all this happens in 2020 or 2025 or 2030 is anyone's guess. Usually, short-term progress on grand projects like this is much slower than one expects, but in the long term, advances can be surprising.

Consider Mars exploration. When the *Cassini* project began in 1990, nothing had been sent to Mars since 1975 except two Russian *Phobos* spacecraft (the only interplanetary spacecraft more massive than *Cassini*), which failed before they returned much data. A couple of years later, *Mars Observer* exploded on arrival in orbit. It was an agonizing wait for Mars scientists as missions in subsequent years slowly started to recover the situation. And yet as of this writing, in 2006, there are no less than four spacecraft orbiting Mars (*Mars Global Surveyor*, *Mars Odyssey*, and the recently arrived *Mars Reconnaissance Orbiter* slowly aerobraking into its mapping orbit, plus the European *Mars Express*) and two rovers, *Spirit* and *Opportunity*, trundling around on the surface. So a decade and one-half is a long time in the space business generally, even if it is only the typical length of a project to the outer solar system.

In the meantime, *Cassini*'s torrent of data will sustain an extended campaign of scientific study for well over a decade after it ends. New generations of graduate students will trawl through the data, applying new methods and approaches. What happens next will depend on the relative importance of different branches of science, and on the politics of space exploration. A technologically risk-averse approach would favor an orbiter, which would address primarily geological and geophysical objectives—global mapping, global topography, gravity, and magnetic fields—as well as monitor the weather in the lower and perhaps upper atmosphere. We know how to do orbiters.

If, on the other hand, astrobiology and organic chemistry are perceived as being important enough, the priority would shift toward in situ exploration, and a balloon able to sample surface material. Such a mission would benefit from an orbiter, but in principle could be accomplished without one.

But there is competition in the outer solar system. Many scientists favor Europa as a scientific target. It is arguably a more relevant astrobiological target, with a water ocean (and perhaps hydrothermal vents like the Earth) closer to the surface than Titan's. However, even that ocean is inaccessible as far as today's technology goes, and airless organic-poor Europa is a body with narrower scientific appeal than Titan.

Another lobby favors a Neptune orbiter. Although all agree that Neptune and its exotic geysering moon Triton, as well as its smaller moons and its rings, deserve systematic exploration much like *Cassini* has done for Saturn, the harsh reality is that reaching Neptune to enter orbit around it takes decades. The dilemma is that, the faster one flies the 30 AU to Neptune, the more effort one needs—as rocket power or in the form of an aerocapture heat shield—to stop when one gets there. *Voyager* reached Neptune in about twelve years but didn't stop. It spent only a meager few weeks close enough to do better than Hubble.

Titan, however, is not getting any less fascinating with time as *Cassini*'s findings draw interest from sedimentologists to study its dunes, from chemists to study its atmosphere and haze, from meteorologists to study its clouds, oceanographers and limnologists to study its lakes. And for engineers, too, Titan has particular appeal. Although NASA is often seen as a scientific enterprise, in many respects it is more a "mission" organization, oriented around organizing and building spacecraft. A Titan mission, involving aerobraking into orbit and depositing a balloon and maybe a lander too, offers a wide range of interesting but tractable technical challenges.

*Cassini*'s discoveries at Saturn have also thrown a new player into the ring: Enceladus. If liquid water is the key to finding life, then Enceladus seems a much more realistic prospect for actually touching the stuff than Europa. Although a Europa orbiter is difficult and expensive because of the rocket power needed and the intense radiation at Jupiter, a lander is tougher still. Yet Enceladus is cooperatively hosing its water (and organics with it) out into space, where it can easily be caught and analyzed.

What the next few years will tell us, and the effects they will have on the direction of future exploration, is impossible to guess. Maybe a Europa mission will go first after all, or maybe a diversion of NASA's resources into sending people to the Moon will stall outer solar system exploration altogether for a while. But sooner or later, we will go back to Titan.

## TITAN: A NEW WORLD

The tale of unveiling Titan in the last year or two has been one of science in action, with some ideas tested by data and found wanting, and in other cases guesses being smugly confirmed. Still others are ideas that we want to be true, but the data so far are not good enough to be sure. It is work in progress.

Whatever happens in the future, the last few years have seen a revolution as far as Titan is concerned. Just as Mars has changed since the late 1960s from a world of mystery, with scientists struggling to interpret a few patterns of bright and dark seen through a telescope, into a world whose landscape and climate we are familiar with, and on which our machines (and perhaps one day ourselves) can explore, so Titan has been transformed into a known world—a world strangely like our own. Like Mars's dust storms, frost cycle, and northern lowlands, Titan's polar hood, methane monsoons, and equatorial sand seas are entering our consciousness as features of the universe around us. And like the Sea of Tranquility on the Moon (the landing site of *Apollo 11*) or Meridiani Planum on Mars (the landing site of the *Opportunity* rover), so Shangri-La, Belet, and Xanadu will become names that instantly evoke pictures—the same pictures for all of us, since we all share the same robot proxy—of a far-off landscape remarkably similar to the ones we live in but nevertheless exotically different.

# Appendix

## TITAN: SUMMARY OF DYNAMICAL AND PHYSICAL DATA

| | |
|---|---|
| Radius (surface) | 2,575 km |
| Mass | $1.346 \times 10^{23}$ kg = 0.022 mass of Earth |
| Mean density | $1,880$ kg/m$^3$ |
| Surface temperature | 94 K = $-179°$C |
| Surface pressure | 1.44 bar |
| Surface gravity | 1.35 m/s$^2$ |
| Escape velocity | 2.65 km/s |
| Bond albedo[1] | 0.29 |
| Magnitude $(V_0)$[2] | 8.3 |
| Mean distance from Saturn | 1.223 million km = 20 Saturn radii |
| Mean distance from Sun | 9.539 AU[3] = 1,427 million km |
| Orbital period around Saturn | 15.945 days |
| Mean orbital velocity | 5.58 km/s |
| Orbital eccentricity | 0.029 |
| Inclination of orbit[4] | 0.33° |
| Obliquity[5] | 26.7° |
| Rotation period | 15.945 days |
| Orbital period around Sun | 29.458 years |

NOTES

[1] The bond albedo is the ratio of total reflected light to total incident light.

[2] $V_0$ is magnitude in visible light at opposition.

[3] AU = astronomical unit. 1 AU = 149,597,870 km, Earth's mean distance from the Sun.

[4] Relative to Saturn's equatorial plane.

[5] The inclination of the equator to the orbital plane.

# Further Reading

The development of the mission, and of our understanding of Titan, is described in our own book, Ralph Lorenz and Jacqueline Mitton, *Lifting Titan's Veil*, Cambridge University Press, 2002.

An earlier, more academic book, which gives particular emphasis to spectroscopy, is Athena Coustenis and Fred Taylor, *Titan: The Earth-Like Moon*, World Scientific, 1999.

An excellent overall description of the *Cassini* mission as a whole is David M. Harland, *Mission to Saturn*, Springer-Praxis, 2003. (A new edition, entitled *Cassini at Saturn*, with more than one hundred additional new pages, was released in 2007.)

A recent management book that derives some lessons from *Cassini* (with some interesting project history and many interviews with *Cassini* managers, scientists, and engineers) is Bram Groen and Charles Hampden-Turner, *The Titans of Saturn*, Marshall Cavendish Business, 2005.

## ACADEMIC READING

To date, most *Cassini* findings have been published in short papers in the journals *Science* and *Nature*. More detailed analyses will be reported in these and in planetary science journals such as *Icarus*, *Journal of Geophysical Research*, and *Planetary and Space Science*. Many papers in these journals describe results from the cruise phase of the mission as well as Earth-based observations and theoretical models. A resource for searching for such papers is http://adsabs.harvard.edu/ads_abstracts.html.

Detailed descriptions of the *Cassini* instruments are given in papers in *Space Science Reviews*, Vol. 104 (2002) and Vols. 114 and 115 (2004). Huygens and its instruments are described in detail in A. Wilson (ed.), *Huygens, Science, Mission and Payload*, ESA SP-1177, 1997.

The earliest compilation of Titan science is D. Hunten (ed.), *The Atmosphere of Titan*, proceedings of a workshop held July 25–27, 1973, at Ames Research Center, NASA SP-340, 1974.

*Voyager* results at Saturn (with a long chapter by Hunten and others devoted to Titan, pp. 671–759) are discussed in the book T. Gehrels and M. S. Matthews (eds.), *Saturn*, University of Arizona Press, Tucson, 1984.

A detailed snapshot of Titan science at the beginning of the *Cassini* project is B. Kaldeich (ed.), *Proceedings of the Symposium on Titan*, Toulouse, France, September 1991, published by the European Space Agency as SP-338, 1992.

Another snapshot, just before *Cassini*'s arrival, which also includes an excellent set of historical papers on Christiaan Huygens and his various activities, is K. Fletcher (ed.), *Proceedings of the International Conference: Titan, From Discovery to Encounter*, April 13–17, 2004, ESTEC, Noordwijk, the Netherlands, published by the European Space Agency as SP-1278.

## ON-LINE RESOURCES

The usual disclaimer must be made that these sites were checked at the time the book was completed, but the Web is an ever-evolving entity, so site addresses and their contents may change or be withdrawn as time passes.

The first points of contact should be the JPL and ESA Web sites, which have large collections of images, movies, documentation, and products for educators, *http://saturn.jpl.nasa.gov* and http://sci.esa.int/huygens.

Many images can be found on the planetary photojournal, http://photojournal .jpl.nasa.gov.

Many informal compilations of *Cassini* results and news can be found on the Web. A particularly complete and well-informed collection is that of the Planetary Society, http://www.planetary.org.

A neat tool for your computer desktop is the Titan Sunclock developed by Mike Allison and colleagues at the Goddard Institute for Space Science, http://www.giss.nasa.gov/tools/titan24/.

The raw and processed science data from the *Cassini* mission can be obtained at NASA's Planetary Data System, http://pdsimg.jpl .nasa.gov/Missions/Cassini_mission.html.

*Huygens* data are available at the ESA Planetary Science Archive, http://
www.rssd.esa.int/index.php?project=PSA&page=index, and mirrored on
the PDS, http://pds-atmospheres.nmsu.edu/data_and_services/
atmospheres_data/Huygens/Huygens.html.
Hubble Space Telescope data are available at http://www.stsci.edu.
The first author's Web site is http://www.lpl.arizona.edu/~rlorenz.

# Index